C000024741

The
Wenner-Gren
Foundation
For Anthropological Research, Inc.

Where the Wild Things Are Now

WENNER-GREN INTERNATIONAL SYMPOSIUM SERIES

Series Editor: Leslie C. Aiello, President, Wenner-Gren Foundation for Anthropological Research, New York.

ISSN: 1475-536X

Previous titles in this series:

Since its inception in 1941, the Wenner-Gren Foundation has convened more than 125 international symposia on pressing issues in anthropology. These symposia affirm the worth of anthropology and its capacity to address the nature of humankind from a wide variety of perspectives. Each symposium brings together participants from around the world, representing different theoretical disciplines and traditions, for a week-long engagement on a specific issue. The Wenner-Gren International Symposium Series was initiated in 2000 to ensure the publication and distribution of the results of the foundation's International Symposium Program.

Prior to this series, some landmark Wenner-Gren volumes include: *Man's Role in Changing the Face of the Earth* (1956), ed. William L. Thomas; *Man the Hunter* (1968), eds Irv DeVore and Richard B. Lee; *Cloth and Human Experience* (1989), eds Jane Schneider and Annette Weiner; and *Tools, Language and Cognition in Human Evolution* (1993), eds Kathleen Gibson and Tim Ingold. Reports on recent symposia and further information can be found on the foundation's website at www.wennergren.org.

The
Wenner-Gren
Foundation
For Anthropological Research, Inc.

Where the Wild Things Are Now

Are Now

Domestication Reconsidered

Edited by

REBECCA CASSIDY AND MOLLY MULLIN

Oxford · New York

First published in 2007 by
Berg
Editorial offices:
1st Floor, Angel Court, 81 St Clements Street, Oxford, OX4 1AW, UK
175 Fifth Avenue, New York, NY 10010, USA

Berg is the imprint of Oxford International Publishers Ltd.

Library of Congress Cataloguing-in-Publication Data

Where the wild things are now : domestication reconsidered / edited by
Rebecca Cassidy and Molly Mullin.
        p. cm. — (Wenner-Gren international symposium series, ISSN
1475-536X)
    Includes bibliographical references and index.
    ISBN-13: 978-1-84520-152-4 (cloth)
    ISBN-10: 1-84520-152-3 (cloth)
    ISBN-13: 978-1-84520-153-1 (pbk.)
    ISBN-10: 1-84520-153-1 (pbk.)
    1. Domestication—Congresses. 2. Domestic animals—Congresses.
3. Plants, Cultivated—Congresses. 4. Human-animal relationships—
Congresses. 5. Human-plant relationships—Congresses. I. Cassidy,
Rebecca. II. Mullin, Molly H., 1960- III. Wenner-Gren Foundation for
Anthropological Research.
    GT5870.W54 2007
    304.5—dc22

                                                                2007001489

British Library Cataloguing-in-Publication Data
A catalogue record for this book is available from the British Library.

ISBN 978 184520 152 4 (Cloth)
ISBN 978 184520 153 1 (Paper)

Typeset by JS Typesetting, Porthcawl, Mid Glamorgan
Printed in the United Kingdom by Biddles Ltd, King's Lynn

www.bergpublishers.com

To Anthony James Cassidy
and
Frances Z. Mullin-Saunders

Contents

Acknowledgments

This volume is the result of a Wenner-Gren Foundation International Symposium, "Where the Wild Things Are Now," which took place in Tucson, Arizona, March 12–18, 2004. We wish to thank the Wenner-Gren Foundation for providing such a congenial setting for us to share ideas and explore areas of common interest. Richard Fox, the foundation's president at the time, responded to our initial proposal for a symposium and to our ideas as they developed with a helpful combination of skepticism and open-mindedness. Sarah Franklin coorganized two sessions at the Annual Meetings of the American Anthropological Association in 2000 and 2001 that were important first steps in the development of this project. She also introduced us to one another at a conference in Bath in 1999 and for that we will always be grateful. Sarah was in Australia at the time of the symposium and unable to attend, but her presence was nonetheless felt and appreciated. We thank Laurie Obbink for incredible organizational skills, enthusiasm, and kindness and Amy Perlow for her able assistance in Tucson. We thank the volume contributors and our two discussants at the symposium, Keith Dobney and Shivi Sivaramakrishnan, for their patience and for their determination to overlook the usual disciplinary boundaries that often separate what, to us, are closely related arguments about the future of the concept of domestication. Finally, we wish to express our appreciation of the Wenner-Gren Foundation's generous support for the production of this volume.

Rebecca Cassidy
Molly Mullin

List of Figures

Participants at the Wenner-Gren Foundation International Symposium "Where the Wild Things Are Now"

Rebecca Cassidy, Goldsmiths College, University of London
Nigel Clark, Open University
Keith Dobney, University of Durham
Gillian Feeley-Harnik, University of Michigan
Agustin Fuentes, University of Notre Dame
Helen Leach, University of Otago
Marianne Lien, University of Oslo
Pamela D. McElwee, Arizona State University
Molly H. Mullin, Albion College
Karen Rader, Virginia Commonwealth University
Nerissa Russell, Cornell University
Shivi Sivaramakrishnan, University of Washington
Yuka Suzuki, Bard College
Peter J. Wilson, University of Otago

Introduction: Domestication Reconsidered

Rebecca Cassidy

The wilderness as a beleaguered space and contested concept has received a great deal of attention in the writings of environmental historians including William Cronon and others (1995; Crosby 1986, 1994; Oelschlaeger 1993; Schama 1995; Worster 1977, 1994). According to Cronon, the idea of wilderness employed by many environmentalists perpetuates distinctions between society and nature, human and animal, and domesticated and wild. Anthropologists have long recognized that these distinctions are anything but obvious and immutable, and that their most powerful incarnations arise from the enlightenment thinking described by Latour (1993). The idea that the authentic wild is somehow "out there" occupying space that is untouched by human influence distorts understandings of places that are not out of time, or out of space. Numerous ethnographies of national parks and other "wild" spaces have shown that they are imbricated within complex national, regional, and international spheres of influence (see, e.g., Brockington 2002; Neumann 1999; Walley 2004; for an excellent overview of this work see Borgerhoff Mulder and Copolillo 2005). The idea of "wilderness" has successfully been complicated in such a way as to put the social back into the wild (MacNaughton and Urry 1998; Whatmore 2002). However, less attention has been paid to the other side of this equation, the domesticated.

In anthropological lore, domestication was an event that took place between 10 and 12 thousand years ago, at which time a transition took place, from savagery to barbarianism (Morgan 1877), private property was created and womankind was universally defeated (Engels

1972/1884). The cause of this event has been identified as a massive growth in population (Cohen 1977; McNeill 2000; Ponting 1992), settled living (Ponting 1992: 37) or climate change (first by Childe 1928, and more recently by Sherratt 1997). Childe, following Engels, described the advent of agriculture as "a revolution" (Childe 1928: 2) and geologist Nathaniel Shaler wrote in 1896 that alongside animal domestication there arose "a civilised state of mind" (Anderson 1997: 467). In many of these older texts, the domestication of plants and animals is seen as a human achievement based on breakthroughs in knowledge that assume similar structures to advances in scientific knowledge, that is, they take the form of experiment, deduction, and repetition. Early farmers are portrayed as agriculturalists in the mold of a Townsend, Bakewell or Coke, pondering the fields and their contents and selecting their raw materials with the aim of ensuring repeatable results. Biological anthropologists and archaeologists have realized that this story of domestication, like that of the "wilderness," is a simplification of a process that is both more partial and more interesting. This collection attempts to impart complexity to the story, in the hope that the "domesticated" will no longer be treated as the unmarked, unproblematic categorical opposition to the complex and powerful notion of the wild.

   The contributors to this collection maintain that uses of the concept of domestication, like those of culture, are helpful in understanding the history of anthropology—the ways in which both "culture" and "domestication" are employed by anthropologists are often indicative of distinctive features of our discipline at particular times. For example, when Juliet Clutton-Brock defined domestication in 1989, she emphasized human control and the conversion of animals into property (p. 7). More recently, scholars, including some archaeologists, but primarily working outside anthropology, have emphasized the mutuality of domestication, by reconstructing the development of relationships between humans and other species of animals and plants. They have deemphasized notions of ownership, property, and control, in favor of a more flexible nexus including cooperation, exchange and serendipity (Pollan 2001). Steven Budiansky has gone even further, crediting animals and plants with the domestication of humans (1992). This argument has a long and venerable history, and definitions of domestication as symbiosis between creatures have long coexisted alongside alternatives that stress human mastery (Hahn 1896; O'Connor 1997; Rindos 1984; Zeuner 1963). This perspective has also been taken up by ethnobotanists, who have described a flexible distinction between wild and domestic plants based on contemporary data gathered among subsistence farmers and indigenous people (Balee 1998, 1999; Nazarea-Sandoval 1995; Padoch and Peluso 1996; Posey 2002). This research, along with that of anthropologists working on plant conservation, has identified porous, culturally variable distinctions between wild and domesticated (Brush and Orlove 1996; Davidson Hunt and Berkes 2003; Etkin 2002).

The contributors to this volume came together for a Wenner-Gren International Symposium, which took place in Tucson in 2004. The chapters that follow reflect discussions that took place during five days of blissful freedom from the usual demands of work and family, for which we were all very grateful. The shared positions that emerged during this time were based on lively discussions and productive disagreements across and within subdisciplines. The intention of the authors in this volume is to submit the concept of domestication to similar kinds of scrutiny as have been applied to the concept of "culture." Our starting point is that it is useful to discuss the fact that both concepts have been productive and provocative. Many anthropologists would agree and argue that both have value and should not be abandoned. However, as the chapters illustrate, the concept of domestication, like that of culture, is slippery and imprecise, those who invoke it often slide between distinctive meanings. Others rely on a common sense interpretation of the term that leaves the assumptions it contains completely untheorized, although no less powerful. Finally, like "culture," "domestication" has come down to us via a problematic history. This introduction revisits the challenges that this history creates for anthropology and introduces the responses offered within the chapters that follow.

We believe that it is useful for anthropologists to engage with biologists regarding their understanding of domestication, and that this can enable us to be more explicit about how the term might be understood in our own work. However, we are not advocating a "domestication" revival. In fact, we are resistant to a number of the more metaphorical uses of the term in anthropology. What does it mean to domesticate desire (Brenner 1998)? Or Europe (Hodder 1990)? Or death (Douglas 1978)? Or mobile telephones (Ling 2001)? What does it mean to describe people as undergoing a process of domestication (Go 1997)? Outside anthropology, domestication is now seen as an appropriate concept with which to analyze, for example, the incorporation of Information and Communication Technologies into the home (Anderson 2002). In these cases, domestication seems an attractive but ultimately unsatisfying euphemism, with a catch-all property that makes critique impossible. In our more limited understanding of the term, although it may be possible to identify a relationship between humans and other organisms in a particular time and place as one of domestication, it is less enlightening to envisage this relationship between humans per se, and an emotion, a continent or mortality.

It is also problematic to attribute any particular qualities to the process, as has historically been the case, when domestication was recast as degeneration or feminization relative to a virile, robust, and masculine (wild) nature. In many of these cases, scholars are able to use this term because they are completely unconcerned with relationships with other animals and the environment. However, it seems that they should at least be aware of the changes that have taken place in the natural sciences and in archaeology. In zoology, the idea of domestication as an asymmetrical relationship between humans and nonhuman others exists alongside more recent versions that emphasize symmetrical relationships of coevolution. Many social anthropologists, historians, and scholars working in related fields continue to use the traditional version of domestication without realizing that this concept has undergone a transformation in cognate disciplines where "domestication" no longer unambiguously denotes a conscious and unequal power relationship between distinct agents.

Domestication can sometimes seem to be taken for granted as a state of affairs the origins of which are now lost in the mists of time, the effects of which will be with us forever. However, this fails to take into account the fact that, in global historical terms, domestication is a relatively recent phenomenon, and in population terms, more than 90 percent of Homo sapiens have lived by hunting and gathering. This calibration helps to undermine some of the definitions of domesticity as the "natural" (ironically) relationship between humans and their environment. By making domesticity the hallmark of humanity, we exclude 90 percent of our ancestors from that category. By including only the kinds of domesticatory relationships that are embodied by selective breeding and other features of post industrial agriculture, we exclude 97 percent of the estimated 80 billion people who have lived (Zubrow 1986). A the symposium that gave rise to this collection, the different temporal and spatial scales, priorities and histories of definitions and uses of the term domestication within subdisciplines encouraged each of us to look at the concept with new eyes.

The aim of this collection is to connect the long history of domestication told within archaeology, and the recent reworkings of this history that take seriously the impact of unintended consequences and the roles of nonhuman participants, to contemporary refashionings of relationships between humans and the environment. On the one hand, domestication has a technical biological definition, on the other hand, it is used as a metaphor within the human sciences including history, social anthropology, and cultural studies. However, in our discussions, we discovered that just as users of the biological definition benefited from considering the dynamic and powerful relationships that social and cultural anthropologists were attempting to capture by using the term, so they in turn benefited from a more rigorous consideration of what it was they actually meant. Interdisciplinarity forced us to make explicit the unspoken assumptions within which each of us operates when using the concept of domestication. It is worth remembering that "domestication" was originally defined as "becoming accustomed to the household" and that the application of this term to plants and animals after Darwin is the first euphemistic use (Bulliet 2005: 7). This is not, therefore, a search for a definition that is somehow more "accurate," but rather a plea for precision and for making explicit the currently hidden implications of usage.

The Use of Concepts of Domestication within Anthropology

Domestication has been a traditional concern within anthropology, particularly within the subfields of archaeology and biological anthropology. It has recently been revived in the subfield of social anthropology, but in the mean time the understanding of domestication within the biological subdisciplines have changed. This section offers an overview of uses of the concept of domestication within anthropology and its transformation outside. The next section introduces the chapters and how they respond to this new challenge.

When Juliet Clutton-Brock edited the last collection to focus on the concept of domestication within anthropology it was defined as a "a cultural and a biological process ... that can only take place when tamed animals are incorporated into the social structure of the human group and become objects of ownership" (1989: 7). Domesticated animals were defined as, "bred in captivity for purposes of economic profit to a human community that maintains complete mastery over its breeding organization of territory and food supply" (1989: 7). The definition was primarily economic and biological and based on the incorporation of animals into exploitative relations with humans. This definition of domestication and the domesticate has coexisted alongside other competing notions of relationships between humans and the environment that emphasize mutuality, fallibility, and chance (Hahn 1896; O'Connor 1997; Rindos 1984; Zeuner 1963). They replace the unidirectional, progressive history of increasingly exploitative relationships with the environment with a more halting and incomplete version, in which the experience of sheep and goats, for example, are placed against the more ambiguous stories of microbes and weeds. In this competing vision, emphasis has been placed on mutual interaction between humans and nonhuman species, and the variety of cultural strategies that are employed in managing human and nonhuman resources, the role of animals and the environment in this negotiation, and the changing balance between exploitation, accommodation, and mutuality.

In archaeology, Shaler's idea of domestication as a strategy undertaken by people who were distinct from their "savage" predecessors has been remarkably tenacious, and in 1984 Bradley observed that, "in the literature as a whole, successful farmers have social relations with one another, while hunter-gatherers have ecological relations with hazelnuts" (Richards 2003: 136). One consequence of the development of archaeology alongside nationalist ideologies was a strategic emphasis on those ancestors who best embodied the traits most valued in the contemporary population. In Europe, this involved the relative neglect of hunting and gathering ancestors in favor of early farmers. The most influential example of this argument appears in the work of Marxist archaeologist V. Gordon Childe, who, in 1925, argued that the Neolithic Revolution in Europe was effected by farmers migrating from the Near East, usurping the existing "simpler" societies of the Mesolithic. Since the mid-1980s this balance has been redressed by studies of molecular genetics that have revealed the contributions of both farmers and foragers to modern populations and undermined assumptions of the "farmers are us" variety. This marks what Richards has described as a "rehabilitation of Europe's hunter-gatherer past" (2003: 136), a task that is inherently interdisciplinary, involving biologists, ecologists, geographers, anthropologists, and archaeologists (Brown 1999).

Contrasting positions on animal domestication within archaeology are perhaps best captured in work on the dog; widely considered to be the first domesticate. In 1997, Robert Wayne used mitochondrial analysis to establish that the most likely ancestor of the dog is the wolf (Vilà et al. 1997). Having settled Konrad's question as to whether dogs were the descendents of jackals or wolves, the question remained as to the mechanism by which a wolf becomes a dog. Biologists including Raymond Coppinger (2002) have argued that the dog domesticated itself, in the sense that the most adept scavengers within the canine population enjoyed an evolutionary advantage over those that adapted less well to this lifestyle, eked out on the rubbish heaps of human society. This theory has usurped the idea that dogs descend from wolf cubs captured from the wild and trained to serve as guards or companions, and replaces a scenario in which people drew the dog into their community for purposes practical or social (Serpell 1989), with a story about opportunism on the part of the animal. The argument is based on DNA-level research that suggests that dogs were domesticated on multiple occasions, and for the first time 100 thousand years ago, rather than the 12- to 14-thousand year estimate based on fossil evidence. The DNA evidence is controversial, however, and the much earlier date of domestication is based on the idea that the changes that appear in the fossil record may have been caused by sedentism, and early dogs may have been morphologically similar to wolves (Vilà et al. 1997). Coppinger, for example, is interested in dogs as companion animals—his wife is a sled dog trainer—however, the purpose of this research, including the rendering of the canine genome, is to provide a model for the inheritance of human disease (Wayne and Oastrander 2005). The secrets that we are discovering about the earliest relationships between dogs and humans are the by-products of a new technique of exploitation, one that epitomizes contemporary relationships between humans and animals, and is conducted in laboratories, via their genes and ours (Haraway 2003).

Social and cultural anthropologists also come to domestication via an intellectual trajectory that is strongly influenced by Marx. Evolutionist anthropologist Lewis Henry Morgan, described by some as the father of U.S. anthropology, spent a considerable period of time working with the Iroquois in the 1840s, and was concerned that aspects of their social life be recorded before they disappeared. He is best known for his work on kinship, and for his distinction between classificatory kinship systems, which do not discriminate between linear and collateral kin, and descriptive kinship systems, which do. In *Primitive Society* (1877), Morgan combined his knowledge of the Iroquois with large quantities of secondary data from elsewhere in the world to argue that all societies progressed through three distinct phases: savagery, barbarism, and civilization (savagery and barbarism each consisted of three further subsections: upper, middle, and lower). These stages were technologically defined and associated with particular family structures and political organization. Middle barbarism was marked by animal domestication and irrigation, civilization by the phonetic alphabet. Marx read Morgan late in life, and died before he was able to incorporate his insights into his writing, a task undertaken by Engels in the *Origin of the Family, Private Property and the State* (1972/1884), which has become the central text of Marxist anthropology.

According to Engels, "the domestication of animals and the breeding of herds ... developed a hitherto unsuspected source of wealth and created entirely new social relations" (1972/1884: 119). Once private property was amassed within the family, men had an interest in ensuring that their male offspring inherit, leading to the overthrow of matriliny and what Engels referred to as the "world historical defeat of the female sex" (1972/1884: 120–121). Feminist scholars have since criticized Engels for underestimating the influence of individual women, and subsequent Marxists for using binary oppositions to account for the subordination of women (nature–culture, reproduction–production, domestic–public) that are Western constructs rather than cultural universals (Harris 1980; Strathern 1980). However, criticism of the putative consequences of domestication for women do not affect the assumption that domestication was deliberate, unidirectional, and of profound significance in the transformation of the basic struct-ures of society. Pierre Ducos (1989) and Tim Ingold (1980) have pur-sued the idea of domestication as becoming property, conditions under which animals and plants may exist for thousands of years before the morphological changes, measured by archaeologists and biological anthropologists and regarded by them as the hallmark of domestication, appear.

Social anthropologists are also interested in exploring the use of concepts of domestication to establish moral hierarchies, along with scholars from geography (Anderson 1997; Whatmore 2002), philosophy (Noske 1989), and science studies (Birke 1995; Haraway 1991; Plumwood 1993). This work examines the concept of domestication as implicitly attributing or denying humanity to various groups on the basis of their relationship with the built environment, plants, and animals. Most obvious and insidious of these distinctions was made at the height of the colonial era, when, in accordance with Shaler's work, those who practiced animal husbandry were classified as "civilised" and those who did not, as "savage." In a different context, domestication became associated with degeneration, and domestic animals in particular were seen as corrupt and inauthentic versions of their wild ancestors. The domestic, when opposed to the wild in this way, became one of a set of binary oppositions including nature–culture, private–public, female–male, animal–human, and child–adult. This argument created "the savage within": categories of people in European society to whom it was not possible to extend full human status, including women, children, and the mentally ill. Critical scholars have exposed the resilience of these implications, of civilization on the one hand and feminization (implying weakness and a lack of virility) on the other hand, and their continued influence if the implications of different understandings of the concept of domestication are not made explicit. The contributors to this volume think that uncritical uses of domestication risk enforcing these distinctions and are interested in the intellectual trajectory that produced various understandings of domestication, and in turn influenced conceptualizations of differences between groups of animals and people (Feeley-Harnik this volume). This volume is also a consideration of how or whether recent efforts to reconsider the process of domestication may be used to expose the mutual support that notions of civilization and human control received from the sociopolitical context in which earlier understandings of domestication arose.

In contrast to social anthropology, biological anthropology has placed greater focus on the impact of domestication on human behavior and human bodies through morphological trends that can be measured, for example, lighter bones and smaller teeth. In general, the derivation from wild type to domestic type involves a canalization, which results in reduced viability in a so-called "wild" setting. Primatology also separates wild or free-ranging groups from those it describes as semifree ranging and human impacted, and focuses on the former to study processes thought to reside in "nature." The effect of this concern with the authentic "wild" population is to represent domestic animals as having been degraded by their contact with humans, and to romanticize the wild in a way that cultural geographers, social anthropologists and environmental historians have shown to be detrimental to understanding the complex multiple connections between humans, other organisms and the built environment. This understanding of the domestic relative to the wild also evokes the metaphorical usage that has entered social anthropology and cultural studies, of feminization, in which feminization is regarded as a weakening often combined with entry into the *domus*. Recent work in primatology, including the contribution to this volume by Agustin Fuentes, argue strongly that the study of impacted populations should be a priority (see also Fuentes, Southern, and Suaryana 2005; Fuentes and Wolfe 2002; Paterson and Wallis 2005).

Commensals, Diseases, and the Unforeseen Consequences of Interaction: Predators, Pathogens, and Symbionts

At least three sets of relationships force a rethinking of definitions of domestication that focus on human mastery over other organisms: diseases, commensals, and the unintended consequences of interactions with other animals and the environment. In *Plagues and Peoples* (1976), William McNeill described the strain placed on "older patterns of biological evolution" by cultural evolution. The role of uncontrolled organisms, including flora, fauna and pathogens, introduced deliberately or accidentally to colonial contexts, has been carefully explored by Crosby (1986), and, in popular science, by Diamond (1997). These works have replaced triumphalist versions of world history written by Europeans, with stories about unintended consequences, and of biological novelty wreaking havoc among vulnerable populations. The outcome of these encounters, the virtual extermination of an entire population in South America, for example, can be redrawn as the result of epidemiology, rather than the usurping of one (inferior) civilization by another. Like the idea of farmers repopulating Europe during the Neolithic by virtue of their superior way of life, thus eliminating hunting and gathering people, the political motivations of this story now seem glaringly obvious.

In a collection that seeks to reevaluate the relationship between nature and culture, social anthropologist Roy Ellen defines domesticates as, "species which owe their current genetic composition to close encounters with human populations which harvest them for food and other products" (1996: 20). However, archaeologists have struggled to accommodate anomalous cases such as mice and sparrows within this category, some excluding them by fiat (Hemmer 1990) and labeling them "commensals," a category of animals that has "nothing to do with domestication" (1990: 1, 178), others preferring to use this category to problematize the category of "domesticate" (Leach 2003, this volume). Commensals raise particular issues for anthropologists because they have clearly changed as a result of living alongside human beings, but these changes are not the results of conscious intervention. O'Connor, in particular, has used the examples of the house sparrow and mouse to argue for the abandonment of the term "domesticates" in favor of the terms "mutualism" and "commensalism" (1997). Helen Leach (2003) has described the changes induced in the mouse and sparrow by involuntary selection of certain traits connected to the built environment and used them to argue that, in these terms, humans are also domesticates (2003; see also Wilson 1988, this volume).

The coevolutionary perspective of Rindos (1984) raises further questions about existing definitions of domestication. Rindos focuses on plants and argues that domestication is best defined as a symbiotic relationship between a particular taxon and the animal that feeds on it. This definition neither requires a human referent (other animals and plants have long established these kinds of relationships) nor an overarching conscious intent: it is a blind process. This argument has also been used to associate particular religious worldviews with domestication, in the work of, for example, Hahn (1896), Sauer (1952), Isaac (1970), and, most recently, Bulliet (2005). These scholars reverse the Marxist logic described in the previous section. Bulliet argues that we have entered a period that is most usefully described as postdomestic, in which people live far away from animals, and "continue to consume animal products in abundance, but ... experience feelings of guilt, shame, and disgust when they think ... about the industrial processes by which domestic animals are rendered into products and how these products come to market" (2005: 3). This attempt to place the history of domestication into some kind of temporal sequence neglects what we feel are significant distinctions between intentional and unintentional selection, the industrial agricultural practices of selective breeding, and recent interventions at the molecular level. An alternative sequence is suggested by Leach (this volume). Leach's sequence captures the dynamic relationship between the wild and the domestic, hunter-gatherers and agriculturalists, food procurement and food production, and foraging and cultivation, rather than distinguishing between discrete states characterized by fixed properties (Harris 1990; Higgs and Jarman 1969; Zeuner 1963).

Connecting Themes

In this introduction, I can neither hope to provide an exhaustive set of connections between the chapters nor capture every aspect of the discussions that we enjoyed in Tucson. However, several themes do seem especially important. First, to ensure that the insights of subdisciplines can be shared, uses of "domestication" must take into account both the technical and metaphorical meanings of the term, as well as the historical discussion of what is at stake when one understanding is preferred over another. Second, domestication is best viewed as an ongoing relationship between people, animals, plants, and the environment. This relationship may be exploitative or mutual, intentional, or serendipitous, it does not preclude reversals, and although it may appear to go through distinct phases, as in Leach's model (this volume), one tendency does not replace another but merely comes to the fore under particular sets of circumstances that emerge from a combination of social and environmental factors. Third, domestication always takes place in a particular space, which will influence and be influenced by the process, and the most productive discussions of domestication combine aspects of the natural and human sciences (Wilson, Feeley-Harnik, Fuentes this volume).

The first four chapters provide a conceptual and theoretical grounding for the rest of the book. The next four consider domestication through the lens of a specific historical or ethnographic engagement. The final three chapters focus on the politics of the wild. Following an overview of the various historical and contemporary uses of the concept of domestication in anthropology, Nerissa Russell suggests that social anthropologists might fruitfully extend their understanding of animal domestication by analyzing it as a kind of kinship. She uses the new kinship theory of Strathern (1992), Carsten (2000), and Franklin and McKinnon (2001) to illuminate the combination of biological and social relationships involved in both domestication and kinship thinking and argues that, like kinship, domestic animals occupy the borderlands between nature and culture. This heuristic draws out a combination of ideas from archaeology and social anthropology, and presents animal domestication as a kind of classification, a social act invoking intimate and unequal relationships between animals and people and between kin. She suggests that this overlap of prominent issues may be most evident in the phenomenon of bridewealth, a process that often implicates both relations between people and also between people and livestock. Russell concludes by arguing that domestication might best be thought of as a loose collection of practices with intentional and unintentional consequences, which may or may not create biological changes in the animals themselves. These practices are best studied in particular times and places, because meanings of terms and significance of domesticatory relationships (like those of kin) will vary through time and space. Paying attention to the paradoxical use of the techniques of domestication in creating wildness observed by other contributors (Suzuki, Mullin, McElwee this volume) will also help to ensure that the distinction between "wild" and "domesticated" remains fluid and unsettled.

Nigel Clark also describes the intimacy that exists between humans, animals, and plants, and the exchange of gifts, albeit unequal, that takes place between them. Using Diprose's concept of, "corporeal generosity" (2002), Clark questions the assumption of dominance and exploitation that the idea of domestication suggests to many social anthropologists, and instead emphasizes mutuality and the uncontrolled slippages that occur when people, animals and things are brought into close proximity. These exchanges need not always be to human advantage, nor are they always consciously willed by plant, animal or human. They produce a risky, but also potentially liberating, openness, and unpredictable consequences, some of which may involve more symmetrical relationships with animals, others of which extract great costs from either or both parties (AIDS, avaian influenza, H5N1, or SARS).

Such unforeseen consequences, good and ill, are also the subject of Leach's chapter, a rethinking of domestication that deemphasizes conscious intervention and human agency, and examines the biological evidence that "unconscious" selection, rather than artificial selection was responsible for the majority of changes observed in plants and animals subsequent to the Neolithic Revolution. On the basis of a wide survey, Leach suggests that the single, undifferentiated period of domestication, between the Neolithic Revolution and the present, be disaggregated, and divided into four stages, beginning with "unconscious" selection, which left no traces on the wild plants and animals to which it was applied. The second stage involves both unintentional selection and nonspecific breeding principles. The third stage involves the formation of breeds through a combination of cross breeding and inbreeding. And the fourth stage involves techniques that take place at the genetic level. Stage one is identified as having taken place between 10- and 14-thousand years ago, stage two between 5.5-thousand years ago and 2.5-thousand years ago; stage three began 300 years ago, and stage four has been with us for approximately 30 years. One of the most important things to remember about Leach's model is that unconscious selection did not end 10-thousand years ago, and, as Clark's chapter (this volume) reminds us, remains an important part of our (stage four) relationship with domesticates today.

As do several of our contributors, Leach sees the evocation of the "domus," or homestead, as an important feature of the concept of domestication, that should be retained (Russell, Leach, Feeley-Harnik, Mullin this volume). All of our contributors are interested in the spaces in which domestication takes place, and how these are conceived and understood (see, in particular, Fuentes for his discussion of Balinese temples, and Rader for the conceptualizations of the laboratory relative to the home in this volume; see also Ingold 1996; Terrell et al. 2003). Every contributor was asked to engage with the physical environment in its widest sense, animate and inanimate. This insistence was partly motivated by what we perceived as an overemphasis on domestication as a process that involves humans and other life forms, and particularly animals. Haraway's work on the cyborg has taught us that the borderlands between humans and inanimate material (particularly machines) is equally as productive for thinking about the work of purification as is the border between humans and animals (Haraway 1991; Latour 1993). Perhaps the strongest focus on this aspect of domestication is found in the chapter by Peter J. Wilson, who maintains that the usual explanatory direction, that agriculture lead to settlement, should be reversed. He argues that it was in fact architecture, and specifically the construction of permanent settlements, that created the conceptual and material circumstances under which domestication of plants and animals might take place, as he writes, "Without architecture there can be no domestication" (Wilson this volume). Wilson describes how the built environment domesticated people by suggesting new possibilities of protection and separation, through which animals and plants might be contained. According to Wilson, architecture provided physical expressions of relationships that are fundamental to the creation of institutions (inside–outside, public–private, "house rules," status, routine, and the passage of time). Architecture thus provided the basis for a change in thought, from unsystematic or "minimalist" thinking to "patterned thinking." Patterned thinking incorporates minimalist thinking, and is distinguished by composing several different considerations at once. Wilson thus sees the process of domestication as important because it offers up evidence of the activities of the human mind, and perhaps even some answers to the "why" questions, when at present we are limited to questions of "how."

Fuentes explores a complex social and ecological relationship between people and macaque monkeys *(Macaca fascicularis)* in the temples of Bali. He is interested in the macaque monkey's ability to exploit a niche created by human practices, and in the mutual benefits that this adaptation appears to involve. Primates and people presently coexist in many regions of the world, primarily in Africa; South, East, and Southeast Asia; and South and Central America. Although many primatologists are interested in what they regard as "wild" populations in these areas, Fuentes, as stated earlier, is more interested in the "in between," and in particular, in primates that are adapting to human spaces. Primates that engage with humans may do so as prey, pets, or social and economic tools (e.g., the coconut gathering monkeys of Thailand). However, Fuentes chooses to focus on the temple macaques, because their entire social and ecological organization can be said to be affected by and affecting the human society that shares the temple space. This form of niche construction is not as developed where an individual primate is killed as food, or tamed as a pet, or trained as a worker. The physical niche provided by the temple space is not, however, the only condition that gives rise to the relationship that Fuentes has observed at 63 sites around the island. The macaque monkeys are privileged by the islanders as a result of the prevalence of the Ramamyana and other aspects of Hinduism that invoke venerable monkey characters. Thus, the spatial niche is complemented by what Fuentes refers to as a "cultural niche," within which macaque monkeys are tolerated and even fed and protected. The unintended consequences of human cultural and spatial organization benefit both the macaque monkeys and the islanders. Monkeys receive food and shelter, islanders gain culturally and economically from their presence: they are welcome visitors to temple sites, and in some cases also attract tourists.

Gillian Feeley-Harnik's chapter extends the theme of cohabitation by combining the structure of human animal relationships with relations of kinship and residence. It investigates the role of commensalisms and reproduction, the relationship between dominance and domestication, and the historical and contemporary analytic value of the concept of domestication. Focusing on Darwin's use of pigeon-breeding experiments to illustrate his theory of evolution, Feeley-Harnik explores the relationship between artificial and natural selection, and the ambiguous role of domesticated animals in natural history. Unlike his contemporaries, Darwin believed that domestic animals were analogous to wild animals, and could therefore provide data that would assist in understanding the process of evolution that was taking place in nature. Competitors believed that domesticated animals were unsuitable models for wild animals because they were both human creations without the ability to survive in the wild and also demonstrated reversion to type rather than variation. Feeley-Harnik takes Darwin's analogy to its logical conclusion by considering the domestication of pigeons as an example of coadaptation to a particular domus or "house-life" (Morgan 1877), the wider context in which this process took place that was ignored by Darwin.

In the world of fanciers, pigeons were celebrated for their fidelity, their evocation of childhood and most importantly, their embodiment of ideals of craftsmanship. Pigeon fanciers of the time were, according to Darwin, "odd little men" who occupied the margins of great cities and whose interest reflected and reinforced their lack of integration. Spitalfields weavers epitomized these qualities, applying their particular skills to both silk and pigeons, and passing both embodied practices onto subsequent generations. The attention they paid to pigeons reflected the place of birds in the political economy of Britain at the time, an aspect of the environment that Darwin neglects in his abstract rendering of pigeon fancying as unconscious selection. Spitalfields weavers produced both pigeons and the next generation in houses and on roof tops that they occupied for a period of 500 years, from the 1330s to the 1860s, one of a number of waves of workers and animals implicated in shifting local and global networks of production and exchange.

The incentive for commercial actors to present recent genetic level interventions in the lives of animals and plants as part of the ancient and ongoing process of domestication (imagined as intentional and associated with predictable outcomes) is explored in Karen Rader's chapter. Mice enjoy a particularly high profile role in the history of concepts of domestication. They complicate the appendage "domestic-ate," by providing an example of an animal that lives in close contact with people despite, rather than because of, their will. This category of animals, that also includes the house sparrow, has found their niche within human settlements, scavenging on the detritus of life since before 4,000 B.C.E.: their Latin name, *mus muculus,* Rader informs us, derives from an ancient Sanskrit word meaning "thief." The refreshing story of the mouse's predation on the human takes a less edifying turn once the mouse is identified as the object of "fancy" before being recast by science as a limitlessly accessible and renewable model of genetic inheritance. The relationship between fancying and experimenting, and the different tropes of science and pet keeping they enact, have continued to ricochet through our relationships with animals, and are evident in discussions of new genetic technologies and their application to both humans and animals. Rader considers the implications of this elision of human and animal in relation to the production of post genomic biosocialities.

Marianne Lien's description of intensive fish farming in Tasmania offers the intriguing opportunity to observe processes that terrestrial species may have undergone 9,000 years ago, among aquatic species today. The whole lifecycle farming of salmon has only been widespread in the past 30 years, when technology developed in Norway was exported to Canada, Chile, Ireland, Great Britain, Australia, and the case in point, Tasmania. Although it is a rapidly expanding form of food production, it has attracted relatively little academic attention. Lien argues that although fish farming is often associated with capture fishing, viewing it through the lens of domestication enables her to identify considerable parallels with terrestrial farming. At the same time, the particular case she describes raises issues that are specific to Tasmania and to its colonial history, which implicates technology, animals, and people, as well as fish, in networks of (unequal) exchange. The chapter destabilizes the presumed distinction between the wild and the domesticated that often informs discussions of terrestrial farming. Farmed Tasmanian salmon constitute an anomalous species (Douglas 1975), neither fully wild (being both contained and non-native) nor fully domesticated (untamed and contained within underwater net pens that cannot insulate wild and domestic from each other in the manner of hedges and fences on land). By producing the wild in the ocean, Tasmanian aquaculture reveals the dependence of existing definitions of domestication on terrestrial activities.

Various technologies associated with domestication, including inter-ventions that take place at the genetic level, may also be used to produce the wild. Yuka Suzuki's chapter explores how wild animals are produced in Zimbabwe, and how the category of the wild is made powerful in the context of nation building by both settlers and war veterans ("warvets"). She argues that the classification of an animal as "wild" or "domestic" is related to the history of particular human animal relationships. In other words, animals and plants are not born wild or domestic, it is people who designate them as such, according to priorities that are not always obvious, consistent, or permanent. The Zimbabwean example illustrates that this classification is often a violent and powerful act, with considerable repercussions for people and animals alike. Quickly dispatching the idea of "wildlife" as an untouched and pristine realm, Suzuki introduces a context in which humans and animals are mutually implicated, and wild animals may be classified as vermin, game, wildlife, or property, each of these categories with their own context specific frame of reference.

Two important political acts set the scene for wildlife production in Zimbabwe. The first was the designation of wild animals as the property of the owners of the land on which they were found, the second was the policy of sustainable utilization. Suzuki's chapter reverses the logic of questions about definitions of domestication by asking not, "are domestic animals always owned," but instead, "under what conditions is it possible for wild animals to be owned?" She explores the roles of animals in the formation of raced and gendered identities: when black warvets cannot distinguish between a bush pig and a lion, white settlers are given the opportunity to laugh at the state that is threatening to eradicate their way of life. Conversely, warvets killed lions during the land invasions that began in 2000 because they symbolized the colonial history of Zimbabwe in the creation of wealth through wildlife production by white settlers. As Wolch and Emel have argued, the treatment of animals by different groups has often been used to establish moral hierarchies (1998). This chapter strongly establishes the political significance of animal classification, and the importance of historical and contextual understandings of human animal relationships where these are both a source of livelihood and at the center of nation-building activities.

Pamela D. McElwee's chapter also focuses on the political consequences of attributions of wildness, in relation to rice production in Vietnam. In a chapter that could be subtitled: "the importance of being charismatic megafauna," McElwee compares the fate of newly discovered mammals the saola, giant muntjack, and warty pig, with the loss of varieties of rice indigenous to Vietnam. The discovery of new mammals in Vietnam prompted a hysterical reaction from wildlife organizations and others in the international community, and resulted in pledges of $100 million in conservation aid. The loss of biodiversity constituted by the replacement of indigenous varieties of rice by hybrids in the Mekong Delta affects the livelihoods of rice farmers in a region that produces 50 percent of the nation's rice, however, it has attracted virtually no attention from conservationists or other NGOs. The lack of attention and resources attracted by a loss of diversity that drastically impacts on the livelihoods of millions and the prosperity of the nation is brought into strict relief by the contrasting response to the discovery of a new "lost world" in Vu Quang. McElwee's fieldwork among international conservationists reveals that there is a hierarchy of wildness that values animals over plants, the exotic over the mundane, the ancient or primitive over the new, and the wild over the domestic.

Attributions of wildness and conceptions of domestication are not only found in contexts in which the wild may be easily imagined or invoked, as in the cases of safari parks in Zimbabwe or archaeological digs in Vietnam. The "Inner Beast" advertising campaign described by Molly H. Mullin (this volume), reveals the extent to which discussions of the wild and domestic are still powerful resources with which to organize contemporary urban life in North America and Europe. Mullin is concerned with the way in which concepts of domestication can be intertwined with other aspects of everyday life and how we should best study notions of the wild that are unstable and context dependent. She argues that the pet food industry has, in part, helped to create a situation in which consumers look for alternative ways of feeding their pets that can produce emotional bonds between animal and human more easily than it can shake off a past (and parts of a present) that are based on very different kinds of human animal relationships.

The central theme of the "Inner Beast" ad campaign, that domestication requires the input of science and of scientists to solve its problems, is not novel. However, what seems to have changed recently is the increasing emphasis placed on findings from the field rather than the laboratory. The decreasing use of findings from laboratory-based studies in pet food advertising may relate to changes in human animal relationships that have also problematized not just the idea of the pet food industry but also the very idea of the pet. Changing ideas about companion animals find their expression in feeding practices, in evidence at the pet food conventions and other gatherings described by Mullin. In some cases, relationships with pets are increasingly modeled on family relationships, and food must therefore be both edible for, and appetizing to, humans. In contrast, advocates of "wild feeding" emphasize the wild origins of companion animals, activating an entirely different set of concerns and orientation toward the pet food industry. Mullin stresses the problems of general statements regarding domestication, amidst the constant reworkings of the term that are going on in our homes and gardens in relation to our animal companions. She warns, like many of our contributors, against the uncritical use of the concept of "domestication" in which meanings of this concept abound, transformed by and transformative of the contexts in which they take shape.

The chapters explore the variety of ways in which domestication has been used by anthropologists. The contributors are also interested in the propensity of the concept to take on a life of its own: to appear bland or descriptive while at the same time containing resources that are capable of forming the basis of potent connections and distinctions between people, animals, and things. Contributors to this volume are neither attempting to offer a new definition of "domestication" nor promoting or condemning its use. Such a conceit would deny the complexity that I have attempted to capture in this introduction. Like all powerful terms, domestication cannot be contained and new definitions risk temporarily concealing nuances that then burst forth unexpectedly. As Mullin's chapter clearly shows, we are aware that the concept is lively, and that its use is not confined to anthropology, or even academia. We find this complexity challenging, rather than off-putting. Like "culture" it is a useful resource for thinking about our relationship with the environment and will continue to be employed in this context. As such, it is an example of how anthropology, like all intellectual endeavors, is a social practice. This recognition makes the efforts of the contributors, to connect uses of domestication with the contexts that produced them and to anticipate how it might be used in the future, all the more compelling.

This volume provides a foundation for some important questions for the future. What are the implications of codomestication? If, as has been argued, humans are, at least to some extent, domesticating themselves, how does this change how we think about domestication, companion species, ourselves, and the environment of which we are a part? This volume offers not only a number of possible frameworks within which such thinking might take place (Leach, Russell, Wilson) but also an indication of what happens, politically, when this conceptual work remains undone (Suzuki, McElwee, Mullin). It is not simply that we must rethink domestication in the light of new technologies but that we must acknowledge what has always been the case—that domestication is an ongoing and unruly relationship, and that failing to appreciate it as such risks confusing what is contingent (and, therefore, open to criticism and change) with what is inevitable (and, therefore, fixed). The ambition of this volume is to combine insights from subdisciplines that operate within very different scales (through both time and space) to unsettle conventional understandings of relationships between people, other animals, and the environment. To do so successfully is to offer the greatest possible endorsement of the value of an anthropological perspective on the imperfectly domesticated and domesticating new and old wild.

References

Anderson, B. 2003. The domestication of information and communication technologies. In *Encyclopedia of community: From the village to the virtual world*, K. Christensen and D. Levinson, eds. London: Sage.

Anderson, K. 1997. A walk on the wild side: A critical geography of domestication. *Progress in Human Geography* 21 (4): 463–485.

Balée, W. 1998. *Advances in historical ecology*. New York: Columbia University Press.

——. 1999. *Footprints of the forest: Ka'apor ethnobotany—the historical ecology of plant utilization by an Amazonian people*. New York: Columbia University.

Birke, L. 1995. *Feminism, animals and science: The naming of the shrew.* Buckingham: Open University Press.

Borgerhoff Mulder, M., and P. Coppolillo. 2005. *Conservation: Linking ecology, economics and culture.* Princeton: Princeton University Press.

Bradley, R. 1984. *The social foundations of prehistoric Britain.* New York: Longman.

Brenner, S. 1998. *The domestication of desire: Women, wealth and modernity in Java.* Princeton: Princeton University Press.

Brockington, D. 2002. *Fortress conservation: The preservation of the Mkomazi Game Reserve, Tanzania.* Bloomington: Indiana University Press.

Brown, T. 1999. How ancient DNA may help in understanding the origin and spread of agriculture. *Philosophical Transactions, Royal Society of London, Biological Sciences* 354 (1379): 89–98.

Brush, S., and B. Orlove. 1996. Anthropology and the conservation of biodiversity. *Annual Review of Anthropology* 25: 329–352.

Budiansky, S. 1992. *The Covenant of the wild: Why animals chose domestication.* New York: William Morrow.

Bulliet, R. 2005. *Hunters, herders, and hamburgers: The past and future of human–animal relationships.* New York: Columbia University Press.

Carsten, J. 2000. *Cultures of relatedness: New approaches to the study of kinship.* Cambridge: Cambridge University Press.

Childe, V. 1928. The most ancient east: The Oriental prelude to European history. London: Kegan Paul.

Clutton-Brock, J., ed. 1989. *The walking larder: Patterns of domestication, pastoralism and predation.* London: Unwin Hyman.

Cohen, M. 1977. *The food crisis in prehistory: Overpopulation and the origins of agriculture.* New Haven, CT: Yale University Press.

Coppinger, R. 2002. *Dogs—a new understanding of cannine origin, behaviour and evolution.* Chicago: University of Chicago Press.

Cronon, W. 1995. *Uncommon ground: Toward reinventing nature.* New York: W. W. Norton.

Crosby, A.1986. *Ecological imperialism: The biological expansion of Europe, 900–1900.* New York: Cambridge University Press.

——. 1994. Germs, seeds and animals: Studies in ecological history. Armonk, NY: M. E. Sharpe.

Davidson Hunt, I., and F. Berkes. 2003. Nature and society through the lens of resilience: Toward a human in eco-system perspective. In *Navigating social-ecological systems: Building resilience for complexity and change,* edited by F. Berkes, J. Colding, and C. Folkes, 53–82. Cambridge: Cambridge University Press.

Diamond, J. 1997. *Guns, germs, and steel: The fates of human societies.* New York: W. W. Norton.

Diprose, R. 2002. *Corporeal generosity: On giving with Nietzsche, Merleau-Ponty, and Levinas.* Albany, NY: State University of New York Press.

Douglas, A. 1978. *The feminization of American culture.* New York: Knopf.

Douglas, M. 1975. *Implicit meanings.* London: Routledge.

Ducos, P. 1989. Defining domestication: A clarification. In *The walking larder: Patterns of domestication, pastoralism and predation,* edited by J. Clutton-Brock, 28–30. London: Unwin Hyman.

Ellen, R., and K. Fukui, eds. 1996. *Redefining nature: Ecology, culture and domestication,* Oxford: Berg.

Engels, F. 1972 [1884]. *Origins of the family, private property and the state.* New York: International Publishers.

Etkin, N. 2002. Local knowledge of biotic diversity and its conservation in rural Hausaland, Northern Nigeria. *Economic Botany* 56 (1): 73–88.

Fuentes, A., M. Southern, and K. Suaryana. 2005. Monkey forests and human landscapes: Is extensive sympatry sustainable for Homo sapiens and Macaca fascicularis in Bali? In *Commensalism and conflict: The primate-human interface,* edited by J. Patterson and J. Wallis, 168–195. Norman, OK: American Society of Primatology Publications.

Fuentes, A., and L. Wolfe, eds. 2002. *Primates face to face: The conservation implications of human and nonhuman primate interconnections.* Cambridge: Cambridge University Press.

Franklin, S., and S. McKinnon. 2001. Relative values: Reconfiguring kinship studies. In *Relative values: Reconfiguring kinship studies,* edited by S. Franklin and S. McKinnon, 1–25. Durham, NC: Duke University Press.

Go, J. 1997. Democracy, domestication, and doubling in the U.S. colonial Philippines. *Political and Legal Anthropology Review* 20: 50–61.

Hahn, E, 1896. *Die Haustiere und ihre Beziehungen zur Wirtschraft des Menchen.* Leipzig: Duncker and Humblot.

Haraway, D. 1991. *Primate visions: Gender, race and nature in the world of modern science.* London: Routledge.

———. 2003. *The companion species manifesto: Dogs, people and significant otherness.* Chicago: Prickly Paradigm Press.

Harris, D. 1990. Settling down and breaking the ground: Rethinking the Neolithic Revolution. In *Twalfde Kroon-Voordracht,* 437–466. Amsterdam: Stichting Nederlands Museum voor Anthropologie en Praehistorie.

Harris, O. 1980. The power of signs: Gender, culture and the wild in the Bolivian Andes. In *Nature, culture and gender,* edited by Carol MacCormack and Marilyn Strathern, eds., 70–94. Cambridge: Cambridge University Press.

Hemmer, H. 1990. *Domestication: The decline of environmental appreciation.* Cambridge: Cambridge University Press.

Higgs, E., and M. Jarman. 1969. The origins of agriculture: A reconsideration. *Antiquity* 43: 31–41.

Hodder, I. 1990. *The domestication of Europe.* Oxford: Oxford University Press.

Ingold, T. 1980. *Hunters, pastoralists, and ranchers: Reindeer economies and their transformations.* Cambridge: Cambridge University Press.

——. 1996. Hunting and gathering as ways of perceiving the environment. In *Redefining nature: Ecology, culture and domestication,* edited by R. Ellen and K. Fukui, 117–155. Oxford: Berg.

Isaac, E. 1970. *Geography of domestication.* Englewood Cliffs, NJ: Prentice-Hall.

Latour, B. 1993. *We have never been modern.* Cambridge, MA: Harvard University Press.

Leach, H. 2003. Human domestication reconsidered. *Current Anthropology* 44 (3): 349–368.

Ling R. 2001. *"It is 'in.' It doesn't matter if you need it or not, just that you have it.": Fashion and the domestication of the mobile telephone among teens in Norway.* Paper presented at the "Il corpo umano tra tecnologie, comunicazione e moda" (The human body between technologies, communication and fashion) conference, Triennale di Milano, Milano, January, 2001.

MacNaughton, P., and J. Urry. 1998. *Contested natures.* London: Sage.

McNeill, J. 2000. *Something new under the sun: An environmental history of the twentieth century world.* New York: W. W. Norton.

McNeill, W. 1976. *Plagues and peoples.* Garden City, NY: Anchor Press.

Morgan, L. H. 1877. *Ancient society.* London: Macmillan.

Nazarea-Sandoval, V. 1995. *Local knowledge and rural development in the Philippines.* New York: Cornell University Press.

Neumann, R. 1999. *Imposing wilderness: Struggles over livelihood and nature preservation in Africa.* Berkeley: University of California Press.

Noske, B. 1989. *Humans and other animals: Beyond the boundaries of anthropology.* London: Pluto Press.

O'Connor, T. 1997. Working at relationships: Another look at animal domestication. *Antiquity* 71 (271): 149–156.

Oelschlaeger, M. 1993. *The idea of wilderness: From prehistory to the age of ecology.* New Haven, CT: Yale University Press.

Padoch, C., and N. Peluso, eds. 1996. *Borneo in transition: People, forests, conservation and development.* Oxford: Oxford University Press.

Paterson, J., and J. Wallis, eds. 2005. *Commensalism and conflict: The human–primate interface,* vol. 4, *Special topics in primatology.* Norman, OK: American Society of Primatologists.

Plumwood, V. 1993. *Feminism and the mastery of nature.* London: Routledge.

Pollan, M. 2001. *The botany of desire: A plant's-eye view of the world.* New York: Random House.

Ponting, C. 1992. *A green history of the world: The environment and the collapse of great civilizations.* New York: St. Martin's Press.

Posey, D. 2002. *Kayapo ethnoecology and culture.* New York: Routledge.

Richards, M. 2003. The Neolithic invasion of Europe. *Annual Review of Anthropology* 32: 135–162.

Rindos, D. 1984. *The origins of agriculture: An evolutionary perspective.* New York: Academic Press.

Sauer, Carl O. 1952. *Agricultural origins and dispersals.* New York: American Geographical Society.

Schama, S. 1995. *Landscape and memory.* New York: Knopf.

Serpell, J. 1989. Pet-keeping and animal domestication: A reappraisal. In *The walking larder: Patterns of domestication, pastoralism and predation,* edited by J. Clutton-Brock, 10–21. London: Unwin Hyman.

Sherratt, A. 1997. Climatic cycles and behavioral revolutions: The emergence of modern humans and the beginning of farming. *Antiquity* 71: 271–287.

Strathern, M. 1980. No nature, no culture: The Hagen case. In *Nature, culture and gender,* edited by C. McCormack and M. Strathern, 174–222. Cambridge: Cambridge University Press.

——. 1992. *Reproducing the future: Essays on anthropology, kinship and the new reproductive technologies.* New York: Routledge.

Terrell, John Edward, John P. Hart, Sibel Barut, Nicoletta Cellinese, Antonio Curet, Tim Denham, Chapurukha M. Kusimba, Kyle Latinis, Rahul Oka, Joel Palka, Mary E. D. Pohl, Kevin O. Pope, Patrick Ryan Williams, Helen Haines, and John E. Staller. 2003. Domesticated landscapes: The subsistence ecology of plant and animal domestication. *Journal of Archaeological Method and Theory* 10 (4): 323–368.

Vilà, C., P. Savolainen, J. E. Maldonado, I. R. Amorim, J. E. Rice, R. L. Honeycutt, K. A. Crandall, J. Lundeberg, and R. K. Wayne. 1997. Multiple and ancient origins of the domestic dog. *Science* 276 (5319): 1687–1689.

Walley, C. 2004. *Rough waters: Nature and development in an East African marine park*. Princeton: Princeton University Press.

Wayne, R., and E. Oastrander. 2005. The canine genome. *Genome Research* 15: 1706–1716.

Whatmore, S. 2002. *Hybrid geographies: Natures, cultures and spaces*. London: Sage.

Wilson, P. 1988. *The domestication of the human species*. New Haven, CT: Yale University Press.

Wolch, J., and J. Emel. 1998. *Animal geographies: Place, politics and identity in the nature-culture borderlands*. London: Verso.

Worster, D. 1977. *Nature's economy: A history of ecological ideas*. Cambridge: Cambridge University Press.

——. 1994. *An unsettled country: Changing landscapes of the American West*. Albuquerque: University of New Mexico Press.

Zeuner, F. 1963. *A history of domesticated animals*. London: Hutchinson.

Zubrow, E. 1986. Up on the prehistoric farm. Review of Rindos, David. The origins of agriculture: An evolutionary perspective. In *Reviews in Anthropology* 13 (3): 210–222.

The Domestication of Anthropology

Nerissa Russell

In July 2004, Bleda Düring called me to the building he was excavating at Çatalhöyük, a large Neolithic site in central Anatolia (Turkey). While working on a burial pit, he encountered a closely grouped set of sheep's feet, pointing straight up. Together, we excavated what turned out to be the first animal burial at Çatalhöyük: a lamb, lying next to a human skeleton but separated from it by the remains of a mat or blanket that lay over the human and under the lamb (Russell and Düring in press). The lamb lay on its side, with its legs pulled awkwardly straight up, as they must have been carefully held while the pit was filled, perhaps to prevent them from falling across the human body. The intimacy and ambivalence evident in this burial makes the human–animal relationship preserved in this grave hard to label. The lamb's young age (ca. 12 months) prevents an assessment of its morphological domestication, but because the vast majority of the sheep at Çatalhöyük are domestic, this one probably came from someone's flock. None of the other numerous sheep at the site ended up intact in a grave. Why would a pet be so carefully held away from its owner? The enigma posed by this strangely positioned animal encapsulates the multiplicity of relations included under the rubric of domestication.

My purpose in this chapter is to explore some of the uses of the concept of "domestication" in anthropology and archaeology, to point out some of the difficulties of the concept, and to suggest how these very ambiguities can provide fertile ground for future work. Although I will discuss a number of different applications of "domestication," I focus mainly on animal domestication. This is both closer to my own expertise and arguably the arena in which the definition of "domestication" has been most problematic.

For the most part, the concept of domestication has been applied within anthropology to the domestication of plants and animals. Decades of discussion have shown that this is not a simple concept, at least as applied to animals. Many anthropologists have found that biological definitions of animal domestication, drawn from other disciplines, inadequately describe the phenomenon of husbandry. They have tended instead to emphasize social–legal (property rights) aspects, or psychological (domination) factors. The form of biological definition that has found the most favor with anthropologists (chiefly archaeologists) is domestication as symbiosis. At present, the major locus of debate is whether domestication is best understood as symbiosis or a change in social relations.

Meanwhile, in recent decades there has been an increasing tendency to use "domestication" in a broader sense, both beyond but especially within anthropology. These applications play on one or the other of two senses of "domestication": as equivalent to "taming" (although most who address animal domestication in fact distinguish between these two concepts), or drawing on the original roots of the word, referring to the house or household (i.e., playing on the dual sense of "domestic"). An exception is "domiculture," which retains the original biologically based sense of domestication as control but applies it to the environment at large rather than individual plant and animal species (Chase 1989). Although rare or absent in anthropology, in other disciplines "domestication" is occasionally also used in a sense opposing "domestic" to "foreign," as in "the domestication of industry."

The concept and study of domestication has already provided fertile ground for communication and collaboration across disciplinary and subdisciplinary boundaries. In general, archaeologists and biologists and animal scientists find common ground in exploring the biological aspects of domestication. Biologists, archaeologists, and biological anthropologists could fruitfully explore further the similarities between the physical changes of domestication and those seen in human evolution, notably neoteny (Coppinger and Smith 1983; Leach 2003). Sociocultural anthropologists and archaeologists have already joined in the consideration of the social and cultural dimensions of animal domestication. This could usefully be pushed further through analysis of animal domestication as a form of kinship. The broadening use of domestication suggests that many disciplines and subdisciplines could benefit from the work on domestication by others. However, it is also necessary to be explicit about the meaning of "domestication" in each case, and to be precise about which aspects of this multifarious concept are invoked. I propose that approaching the topic in terms of the specific practices of domestication at work in each case will aid in bringing together these various approaches.

The Anthropology of Domestication: A Very Brief History

Although there are earlier discussions, serious studies of plant and animal domestication begin to appear in the second half of the 19th century (e.g., Candolle 1885; Darwin 1868; Galton 1865; Geoffroy Saint-Hilaire 1861; Hahn 1896; Roth 1887). At that point they are based on speculation, textual sources that we now know to be many millennia too late to be useful, and observations of contemporary plants and animals. These accounts tend to be cast in terms of glorious human progress, with humans consciously taking control of animal species. After proposing that plants were domesticated by women, Mason (1966/1895: 258–260), in a chapter entitled *War on the Animal Kingdom,* gives the following account of animal domestication (condensed here):

> In his contact with the animal kingdom, the primitive man developed both militancy and industrialism. He occupies two attitudes in the view of the student, that of a slayer, and that of a captor and tamer. ... It is important to ask how our species came to be masters of the brute kingdom, and what intellectual advantages were gained in the struggle. ... By and by they turned the artillery of Nature on herself. The dog raised a flag of truce and came in to join the hosts of man against the rest. The mountain sheep and the wild goat descended from their rocky fortresses, gave up the contest, and surrendered skins and fleece and flesh and milk to clothe and feed the inventor of the fatal arrow. ... Those that refused to enter in any way into these stipulations are doomed sooner or later to extinction, and many species have already disappeared or withdrawn to the waste places of earth in despair.

With the exception of Pumpelly's (1908) innovative work at Anau in the early 20th century (also now known to involve societies much later than the earliest farmers), there was little archaeological evidence to address plant and animal domestication until after 1950. Notably, the work of Braidwood (Braidwood and Howe 1960) in the Near East and MacNeish (Byers 1967) in Mesoamerica began to provide some direct information on the processes of domestication. The need to identify the beginnings of plant and animal domestication stimulated attention to the nature of this process, first by natural scientists collaborating with archaeologists (e.g., Bökönyi 1969; Ducos 1969; Harlan 1967; Mangelsdorf 1958; Reed 1961; Zeuner 1963), and later by archaeologists specializing in the study of plant and animal remains (e.g., Flannery 1969; Hecker 1982; Jarman 1977; Meadow 1983). Probably in response to these discussions, sociocultural anthropologists have also offered perspectives on the nature and origin of domestication (e.g., Alexander 1969; Haudricourt 1962; Ingold 1980; Sigaut 1980).

Definitions

I have discussed varying definitions of animal domestication and their implications elsewhere (Russell 2002), and will only summarize briefly here. Domestication is difficult to pin down, partly because it involves both biological processes of alteration to organisms and social and cultural changes in both humans and animals. Additionally, a wide range of human–animal relationships is included, at least by some, in the rubric of domestication. It is not easy to find a meaningful definition that includes barnyard animals, ranched livestock, pets, animals such as honeybees (but also pigs in some New Guinea societies) that until recently were routinely captured from the wild and never bred in captivity, tuna confined outside the Straits of Gibraltar and fed with herring imported from the North Sea (Bestor 2001), laboratory mice (Rader this volume), and urban pigeons (Feeley-Harnik this volume) yet excludes wildlife culled or even managed through birth control, restored through captive breeding programs, or held and often bred in zoos (Suzuki this volume). Some of these problems arise through casting definitions in dichotomous terms: "wild" or "domestic." Although some (e.g., Jarman 1977) solve this by arranging human–animal relationships in a continuum of control from random predation to factory farming (or today direct genetic manipulation), I find it more fruitful to think in terms of a spectrum of different *kinds* of relationships. Pets, for example, can be members of either wild or domestic species and represent a fundamentally different relationship from livestock.

Definitions of domestication generally stress either the biological or the social aspects. Those from zoological and animal science backgrounds (e.g., Bökönyi 1989; Clutton-Brock 1994) have tended to stress control of breeding, in particular, along with control of feeding and movement. It is control of breeding that produces biological changes observed in domestic organisms. It must be stressed that control of breeding does not necessarily mean deliberate selective breeding, although this tends to be viewed as the fullest expression of domestication, but in essence refers to practices that lead to genetic isolation from the wild population. Indeed, recent work increasingly tends to view early changes in plants and animals as adaptations to the conditions of cultivation and herding rather than the results of human selection (e.g., Crockford 2000; Harlan 1995: 30–39; Hillman and Davies 1999; Price 2002: 10–11; Zohary, Tchernov, and Horwitz 1998). Given this, some have sought to remove intentionality entirely from the domestication process, characterizing it as an instance of symbiosis or coevolution (e.g., Leach this volume; O'Connor 1997; Rindos 1984). Although this term has not yet entered the archaeological literature, the currently popular ecological concept of "facilitation" may be even more appropriate (Fuentes this volume).

Although coevolutionary models are embraced by many archaeologists, other archaeologists and anthropologists, with their focus on the human end of domestication, have emphasized the social aspects in their definitions. Here, the transformation of animals into property is generally seen as the key to domestication, and as initiating profound changes in both human–human and human–animal relations (e.g., Ducos 1978; Ingold 1980). Suzuki's (this volume) proposal that designated game animals are domestic in the eyes of the state (because they are appropriated as property) while wild in the eyes of the individual takes this in a fruitful direction that helps to resolve some of the difficulties of a simplistic wild–domestic dichotomy. Some sociocultural anthropologists have examined the psychological underpinnings of domestication (e.g., Digard 1990), stressing human domination as the motivating factor for animal domestication, an argument that has been taken up by some archaeologists (Cauvin 1994; Hodder 1987).

Domestication Beyond Plants and Animals

"Domestication" is thus a multifarious concept as applied to plants and animals, where "domestic" is opposed to "wild." It becomes more so thanks to the trend in the last 25 years or so to apply the term to other spheres of human activity, particularly within anthropology. The pioneer in this regard appears to be Jack Goody (1977), whose title *The Domestication of the Savage Mind* plays off Lévi-Strauss's (1966) *The Savage Mind* to cast writing metaphorically as a taming influence, transforming human thought. This sense of domestication as metaphorical taming is followed by other authors. Interestingly, those who study animal domestication generally distinguish between taming and domestication, with "taming" referring to a relationship between an individual animal and an individual human whereas "domestication" involves populations and successive generations. This distinction goes back at least as far as the mid–19th century (Geoffroy Saint-Hilaire 1861). Taming is a necessary precursor to domestication but does not inevitably lead to it. However, when applied to aspects of human life, this distinction is lost.

Wilson (1988, this volume) played on the major significance attached to plant and animal domestication (the Neolithic Revolution) in the evolution of human societies (domestic as not wild), while primarily referring back to the etymological roots of "domestication" (*domus*, the house or household in Latin), in arguing that substantial architecture was more important than the origins of agriculture for human societies in *The Domestication of the Human Species* (domestic as not public).

Outside of anthropology, domestication applied to humans almost always harkens back to an earlier meaning: "to habituate to home life" (Webster 1913: 444) and to feminize (containing senses of domestic as not public, not male, and not wild or out of control). This usage has boomed beginning in the 1990s. Third-wave feminists are reclaiming this older meaning as they explore gender issues in various contexts, usually prior to or outside of contemporary Euro-American culture. Others, though, not writing from a feminist perspective, use "domestication/domesticating" in a pejorative sense, equating feminization with a lessening of vitality (e.g., Placher 1996).

Hodder (1990) manages to draw on virtually all the meanings of domestication in his *The Domestication of Europe*. The term refers most directly to the importance of the house and the household sphere (his *domus*, as opposed to *agrios*) in Neolithic Europe. However, he also builds an argument that the importance of the *domus* lies in its power to domesticate people, more precisely to control the wild in men by bringing the wild into the domesticating female sphere of house and hearth (although he also suggests there may be attempts to control the wild and dangerous in women). Moreover, he follows Cauvin (1972) in arguing that this symbolic domestication of the wild precedes the domestication of plants or at least animals in the conventional sense. Indeed, he suggests that plant and animal domestication may be a by-product of the changes in attitude toward nature brought about by the symbolic control of the wild. Hodder's *domus* model has been extensively critiqued on various grounds (e.g., Tringham 1991; Davis 1992), notably by some suggesting that the obsession with controlling the wild may be more of a modern than a Neolithic concern (Halstead 1996), but has become an established concept in discussions of the European Neolithic and beyond.

Domesticating Kinship

Although there are various ways in which domestication can and does serve as a locus for interaction among disciplines and subdisciplines, here I want to explore what archaeologists and sociocultural anthropologists can gain from thinking about domestication in the context of kinship, and particularly in terms of the "new kinship" theory (e.g., Franklin and McKinnon 2001b; Haraway 1997; Strathern 1992b). At the least, domestication and kinship involve a number of similar issues, some of which I explore here. It may even be appropriate to view animal domestication as an extension of kinship to other species. After all, the term itself implies bringing animals into the household.

Society and Biology

I have briefly alluded to the difficulties and potential rewards of dealing with a concept such as domestication that combines biological and social components. These thoughts are echoed by new kinship theorists. Kinship has traditionally been seen as a social structure built on the foundation of biological relatedness, as gender is to sex. As Franklin (2001) notes, not only do many contemporary anthropologists question the epiphenomenal character of kinship but biology is no longer seen as a solid base. The biological is also constructed, in part modeled on society.

New kinship theory draws on at least two sources in challenging classical approaches to kinship. One impetus is queer theory, which asserts the validity of other kinds of families and relationships than those classically composing kinship structures. Another comes from science studies, in which scholars contemplate the implications of new biological technologies. In vitro fertilization, cloning, and transgenic chimeras create new kinds of relationships that somehow need to be incorporated in reckoning kinship. Theorists such as Donna Haraway (1997) embrace this indeterminacy. Strathern notes that

> kinship [is] a hybrid of different elements. Human kinship is regarded as a fact of society rooted in facts of nature. Persons we recognise as kin divide into those related by blood and those related by marriage, that is, the outcome of or in prospect of procreation. However, the process of procreation as such is seen as belonging not to the domain of society but to the domain of nature. *Kinship thus connects the two domains.* (1992b: 16–17)

Domestication likewise connects society and nature, and we must grapple with but also celebrate its indeterminacy. It is irreducibly both biological and social, and one cannot meaningfully be subordinated to the other. Definitive definitions are impossible, but there are many avenues to explore.

Taxonomy

Kinship is of course a classificatory system that orders human beings. Domestication itself involves a simple (on the surface) classification into wild versus domestic. It complicates biological taxonomy, however. Biological taxonomy is kinship writ large, defining relationships among species. Linnaean taxonomy is a kinship system based only on descent; it, too, is based on notions of degrees of relationship. There has been considerable controversy about how to classify domestic animals (e.g., Corbet and Clutton-Brock 1984; Gautier 1993, 1997; Gentry, Clutton-Brock, and Groves 1996; Groves 1995; Uerpmann 1993). Should domesticated animals be considered separate species from their wild ancestors? Are they merely subspecies? Do we need a separate category *(forma)*? Should the same taxonomic criteria apply, for example, to dogs, which when feral no longer revert to wild type as most domestic animals do, and to reindeer, which are little different from the wild form? What of pigs, whose domestic form is visually quite different from wild boar, but which easily revert to the wild form? As an archaeologist, how do I refer to early domestic animals, whose bones are indistinguishable from their wild ancestors yet were herded in a domesticatory relationship and whose descendents will demonstrate physical changes? Domestication poses challenges to taxonomy that anticipate those of genetic engineering.

Biological taxonomy (folk or Linnaean) has long been used as an analogy to kinship in a variety of ways. The totemic use of plant and animal taxa to classify human groups is a widespread and no doubt ancient example. In a classic article, Leach (1964) famously equates animal edibility on the basis of a classification of human–animal relationships somewhat more complex than wild–domestic with human kinship categories and incest and marriage rules. Thus, pets are like siblings and taboo to eat just as siblings are taboo for sexual relations, and so on. It is not only their proximity that makes pets taboo, though. By becoming virtual family members, they take on quasi-human status and blur human–animal boundaries, so that eating them feels like cannibalism (Fiddes 1991: 133; Shell 1986). Kinship is all about boundaries, both fixing them and crossing them (Franklin and McKinnon 2001a: 19). Not only pets but all domestic animals complicate the boundaries between humans and animals, nature and culture.

Intimacy

The varying degrees of intimacy with animals that are created by domestication pose problems that are solved by mechanisms familiar from kinship systems. I have already alluded to pets (not necessarily domesticated by many definitions because this is essentially a relationship between individual humans and animals that may not lead to genetic isolation), which become virtual humans. Herded animals are not quite so close, but more in the human sphere than are wild animals. Leach (1964) equates them to first cousins and clan sisters: kin with whom marriage is usually prohibited, but extramarital sex may be tolerated. By analogy, he suggests that these animals are edible only if immature or castrated, whereas wild animals (equated to neighbors, friends, enemies, and the pool of potential spouses) can be eaten as intact adults. I am inclined to interpret castration and the consumption of young animals as related to the practical exigencies of herding and the market. However, one could argue that people often overcome the guilt of slaughtering and eating animals they have raised through classifying them as not only less than human, but inferior to their wild counterparts.

Another way of deflecting guilt is to frame slaughter as sacrifice. Prior to sacrifice, the animal is often taunted and tormented, to distance it from its human companions emotionally and symbolically (Serpell 1986: 148–149). This tactic bears a striking resemblance to the joking relationships that mark potentially difficult kin relationships. Such relationships are affinal, involving the crossing of boundaries just as does domestication.

Care, in both the emotional and the practical sense, is an essential part of both kin relations and animal husbandry. Borneman (1997) proposes foregrounding care and caring as a central aspect of kin relations. Clark (this volume) proposes a similar approach to animal domestication.

Power and Property

In addition to classification, "kinship is also utilized to articulate the possibilities for social relations of equality, hierarchy, amity, ambivalence, and violence" (Franklin and McKinnon 2001a: 15). Kinship relations define lines of power within the household and beyond, and specify how property is transmitted. They shape gender and age hierarchies. Kinship systems can do this because "kinship is a technology for producing the material and semiotic effect of natural relationship, of shared kind" (Haraway 1997: 53).

For example, Delaney (1991, 2001) has shown how an understanding of reproduction as involving a male seed containing the essence of the child that is merely incubated in the female has profound implications for kinship and gender relations. Because only the father is truly related to the children, naturally a patrilineal system results. Fathers have property rights in their children, whereas mothers do not, as exemplified in the story of Abraham and Isaac. Women's contribution is minimized and their position marginalized.

Animal domestication inevitably involves domination as well as caring. From the point of view of the animals, many researchers have noted that, with a few minor exceptions such as cats, domesticable animals are those that live in groups with hierarchical dominance systems, among other traits (Clutton-Brock 1994). Humans are able to control them by substituting for the dominant animal. From the human point of view, animals join particular households. In societies in which social relationships are largely conducted within a kinship idiom, it is, thus, highly likely that such animals will be given a status analogous to some kin category.

Livestock often play a key role as bridewealth, and this value may have encouraged the spread of herding (Russell 1998). Here, they are structurally equivalent to women in the kinship system, facilitating marriage and especially paternity. In some African societies in which most marriages are transacted with bridewealth, marriages are also possible without bridewealth, but the children then remain affiliated with their mother's lineage whereas offspring of marriages in which bridewealth has been paid become members of their father's lineage (Goody 1973).

Domestic animals have a complex relationship with power relations mediated through the kinship system. On the one hand, human domination of animals is most likely based in the first instance on prior unequal relations among humans within the household (Ingold 1987: 254). If, however, domestic animals take the position of children, they convert a temporary position of inferiority to a permanent one. On the other hand, domestic animals promote inequality among humans through the patron–client relations they enable (often enacted within the household through the idiom of kinship). Bridewealth, though, tends to be a leveling mechanism across households, so that livestock keep circulating. On a metaphorical level, the extension of kinship categories to animals probably helped enable what would otherwise have been unthinkable: human domination of animals. Foragers generally regard animals as their equals (Ingold 1994). And once domestic animals existed in their permanent subordinate position within the human sphere, it became easier to conceive of humans subordinated like animals. Algaze (2001) traces the emergence of the world's first states in southern Mesopotamia to what he calls the Labor Revolution, based on applying human relations with livestock to subordinate humans.

> Southern elites came to view and use fully encumbered laborers in the same exploitative way that human societies, over the immediately preceding millennia, had viewed and used the labor of domestic animals. This represents a new paradigm of the nature of social relations in human societies. ... Scribal summaries detailing the composition of groups of foreign and native-born captives used as laborers describe them with age and sex categories identical to those used to describe state-owned cattle. ... It would appear that the two classes of labor (captive "others" and domestic animals) were considered equivalent in the minds of Uruk scribes and in the eyes of the institutions that employed them. Early Near Eastern villagers domesticated plants and animals. Uruk urban institutions, in turn, domesticated humans. (Algaze 2001: 212)

Regulating Reproduction

Control of reproduction is both a particular focus of the power relations of kinship and central to most definitions of domestication. This can play out in many different ways, as there are many kinship systems and many styles of herding. Again, human kinship and domestic animals are often intertwined. Domestic animals may affect the form of human kinship. A recent study shows that African matrilineal societies are more likely to become patrilineal if they acquire cattle, and that once established, the combination of cattle and patriliny is very stable, because bridewealth encourages the channeling of wealth through sons (Holden and Mace 2003). More speculatively, Tapper (1988) suggests that cattle herders may tend to be patrilineal to distinguish themselves from their herds, which are intrinsically matrilineal (calves stay with their mothers).

In recent centuries, the careful selective breeding of animals (Leach's eugenic stage [this volume]) has led to the emergence of the pedigree along with formal breeds. As Leach's terminology implies, these notions of "good breeding" and pedigree have been applied to both humans and animals. It is primarily the high status animals (horses, dogs, and to some extent cattle and cats) that have pedigrees, to match those of their owners. Because the stakes are high, there is often some manipulation of pedigrees (false claims of noble ancestry) in both cases (Borneman 1988; Cassidy 2002).

Concluding Ruminations

Despite its difficulties, domestication has been a useful concept for anthropology and archaeology. The difference between appropriating wild resources held in common and ownership of land and animals by individuals or households is a crucial one with major implications for social relations. I have explored briefly a few of these with respect to animal domestication, but there are many more important topics that could be pursued, such as the implications of wealth in land and in animals. Moreover, domestication has recently been applied more widely to other aspects of human life. Some of these uses are tied to plant and animal domestication, and some are not. We can fruitfully consider more deeply the connections among these various domestications, material and metaphorical (and the two are not always easily separated).

In this chapter, I have examined some of the intersections between domestication and another of anthropology's problematic concepts: kinship. I have suggested, on the one hand, that the two concepts encounter similar difficulties, and, on the other hand, that domestication can be approached as a form of kin relation. I now consider each of these propositions a bit further.

Kinship and domestication share some characteristics. Both involve systems of classification, which are in turn tied to issues of power and of affect. Both are transacted primarily in the household sphere broadly construed (the *domus*). Just as human social relations are mediated by kinship structures, so are they negotiated through domestic animals (through exchange of animals, patron–client relations based on loans of livestock, etc.). The two intersect in bridewealth, which is classically transacted mainly with livestock. Both kinship and domestication are inherently simultaneously biological and social. This makes them complicated to study, but the struggle to come to grips with how these aspects interact and shape each other (not one determining the other) is salutary. Perhaps it can even help us to a fuller understanding of both biology and culture.

Strathern (1992a) and Franklin (2003) have suggested that kinship involves not only relationships among people but the creation of what they call "merographic connections" between parts of different spheres, such as biology and society. Certainly domestication also creates these connections, for instance between nature and culture through the bodies of domestic animals. Exploring the nature of these connections will bring us to richer understandings of both kinship and domestication.

However, I am not suggesting that kinship and domestication can be treated as equivalent or fully comparable. Kinship systems involve more complex classification systems than does domestication per se. Kinship can perhaps be summarized as being about procreation and the regulation of interpersonal relations. Domestication is about the transformation of animals both bodily and socially.

Although I think it may be useful to approach animal domestication as a form of kin relation, I would not argue that this is the only way to think of it. In terms of human relations, it seems to me that the most crucial thing about animal domestication is that "wild" animals are converted to property. However, domestication, as its etymological roots imply, also involves bringing animals into the household. Perhaps the transspecies "family" thus created foreshadows the complicated kin (and taxonomic) relations engendered by transgenic creatures. Moreover, animal domestication disrupts the animals' own "kin relations." Tani (1996) stresses that herding depends on practices that disrupt the vertical bond between mother and offspring, and substitute a horizontal bond among same-age animals to form a cohesive herd. Tani also notes that this, in turn, depends on the prior establishment of intimacy between the human herder and the herd.

Kinship may be a useful model for thinking about domestication in another way. It has long been recognized that kinship relations serve many simultaneous purposes that, although sometimes contradictory, are not mutually exclusive. Kinship is an idiom that can be deployed to many ends. The same is true of human–animal relationships, and perhaps this can help us to move beyond the simple wild–domestic distinction that causes us such difficulty. "We do not feel forced in the social world—for example in the field of our relations with kin—to choose between either exploiting others for personal profit or avoiding all direct contact. Yet in the context of relations with animals, this is precisely the choice that is forced on us by the conventional dichotomy between wildness and domestication" (Ingold 1994: 11–12).

Bridewealth is perhaps the most obvious, but certainly not the only place where kinship and animal domestication interact. It has usually been analyzed in terms of "classic" kinship studies. How can we rethink bridewealth in light of the "new" kinship theory? What are the effects of bringing animals into the family—on the animals, the kinship system, and the other members of the household?

The same might be asked of domestic animals more generally. For the most part, the incorporation of domestic animals into human societies that had previously only interacted with "wild" animals must be approached archaeologically. Here, we must expand our studies of animal domestication to include not only the animal bones themselves but the archaeological and social contexts in which they are embedded. I suspect that such studies will reveal not a single process of animal domestication, but multiple kinds of domestication. Some may be more coevolutionary, some more consciously directed to property or provision of meat for diet or sacrifice, and some more concerned with controlling the symbolic power of the animal.

In general, I would propose that we can best bring together the varied work on domestication in anthropology and other disciplines by focusing on domesticatory practices. Rather than worrying whether a particular case "counts" as domestication, we can examine the specific practices used to create domestication or wildness (sometimes simultaneously, as seen in several of the articles in this volume). This focus on practices will help to elucidate the various domestications: in what specific ways are they similar or different? In this way we can build a nuanced understanding of the domesticatory process grounded in local and particular histories. We can link the biological and the social, and the various scales of analysis evident in the studies in this volume.

We may find that some domesticatory practices do not lead to animal domestication in either the biological or the social sense. This is one way to understand the "wildlife farms" analyzed by Suzuki (this volume). Likewise, Fuentes (this volume) shows that feeding temple monkeys has biological and social effects, but it is not clear that it will ever lead to morphological change in the monkeys nor to anyone owning them.

This approach also helps to resolve questions of intentionality. Both intentional and unintentional practices can and often do have unintended consequences. Thus, for instance, intentional herding practices create the conditions that produce unintended alterations in animal and human bodies and behavior (Leach this volume; Zohary, Tchernov, and Horwitz 1998). Moreover, the specific herding practices will affect what changes result, even if these are not the goals of the practices. Or the alteration of the environment through architectural practices intended to provide shelter, privacy, storage, and so on unintentionally create new niches for human–animal interaction that may lead to animal domestication under some definitions, and certainly bodily and behavioral changes (Wilson, Fuentes in this volume).

Elucidating the practices of "wilding" may be an especially useful contribution that anthropology can make to the larger world, as seen in several of the articles in this volume. Lawrence's (1982) examination of rodeo and Morales Muñiz and Morales Muñiz's (1995) of the Spanish bullfight are other excellent studies along these lines. On a more applied level, naive beliefs about wildness lead to poor wildlife management decisions. Here both zooarchaeological studies that clarify the actual history of "wild" animals (e.g., Lauwerier and Plug 2003; Lyman 1996; Reitz 2004) and critical studies of the conceptualization of wildness can be most salutary. Although this is related to the more general literature on the construction of nature, I propose a particular focus on (1) the creation of the wild in the midst of the domestic and in highly structured environments (e.g., Mullin this volume), and (2) the paradoxical use of the very techniques of domestication to produce the wild (e.g., Morales Muñiz and Morales Muñiz 1995).

Domestication is a concept with a long history. Its increasing application to new contexts shows that it retains its power as a way to think about a variety of transformations in human and human–animal or human–plant relations. Rather than trying to choose among ways to define domestication, or to recognize it, we will understand it better by elucidating its practices.

Let us consider the domesticatory practices involved in the human and lamb burial at Çatalhöyük described earlier. We archaeologists immediately wondered whether the lamb was morphologically domestic, a member of the numerous flocks that were tended around the site. This seemed important because Çatalhöyük has yielded a great deal of animal imagery, nearly all of it depicting wild animals. Yet most of the animal remains at the site are of domestic sheep, the main meat source. This lamb was too young to be sure whether it was born into the sphere of daily life (domesticity) or the wild as the inhabitants of Çatalhöyük understood it. However, we can see other domesticatory practices at work. The lamb was quite literally brought into the domestic sphere, buried as humans are beneath a house floor. It was placed with the ancestors and perhaps became an ancestor itself. This seems a strong statement of kinship with the man buried beside it, the three people later buried above this man, and the people who continued to live in the house above. And yet the ambivalence of its position, held awkwardly apart from the man it was buried with and carefully avoided by the subsequent burials, suggests that this kinship was not uncontested. Whatever the nature of the relationship between the man and the lamb, it was a relationship between individuals that crosscut, perhaps uncomfortably, the relations between people and animals, wild and domestic, at Çatalhöyük more generally. The act of burial in itself naturalizes this relationship, just as Haraway (1997) argues that kinship naturalizes relationships, and is itself a domesticatory practice. This small example serves to illustrate how much more we can learn by exploring domesticatory practices than simply by asking the question: Is this domestication?

Notes

This article has benefited immensely from the input of all the participants in the "Where the Wild Things Are Now" symposium. I am grateful to each of them and especially to the organizers, Molly Mullin and Rebecca Cassidy, and to the Wenner-Gren Foundation and its staff that made it all possible. I would also like to thank my colleague, Annelise Riles, who first pointed out to me the similarities between domestication and kinship.

References

Alexander, John. 1969. The indirect evidence for domestication. In *The domestication and exploitation of plants and animals,* edited by P. J. Ucko and G. W. Dimbleby, 123–129. London: Duckworth.

Algaze, Guillermo. 2001. Initial social complexity in southwestern Asia: The Mesopotamian advantage. *Current Anthropology* 42 (2): 199–233.

Bestor, Theodore C. 2001. Supply-side sushi: Commodity, market, and the global city. *American Anthropologist* 103 (1): 76–95.

Bökönyi, Sándor. 1969. Archaeological problems and methods of recognizing animal domestication. In *The domestication and exploitation of plants and animals,* edited by P. J. Ucko and G. W. Dimbleby, 219–229. London: Duckworth.

——. 1989. Definitions of animal domestication. In *The walking larder: Patterns of domestication, pastoralism and predation,* edited by J. Clutton-Brock, 22–27. London: Unwin Hyman.

Borneman, John. 1988. Race, ethnicity, species, breed: Totemism and horse-breed classification in America. *Comparative Studies in Society and History* 30 (1): 25–51.

——. 1997. Caring and being cared for: Displacing marriage, kinship, gender and sexuality. *International Social Science Journal* 49 (154): 573–584.

Braidwood, Robert J., and Bruce Howe. 1960. *Prehistoric investigations in Iraqi Kurdistan.* Studies in Ancient Oriental Civilization, 31. The Oriental Institute of the University of Chicago. Chicago: University of Chicago Press.

Byers, Douglas S., ed. 1967. *The prehistory of the Tehuacan Valley.* Austin: University of Texas Press.

Candolle, Alphonse de. 1885. *Origin of cultivated plants.* New York: D. Appleton.

Cassidy, Rebecca. 2002. *The sport of kings: Kinship, class, and thoroughbred breeding in Newmarket.* Cambridge: Cambridge University Press.

Cauvin, Jacques. 1972. Religions Néolithiques de Syro-Palestine. In *Publications du Centre de Recherches d'Écologie et de Préhistoire.* Paris: J. Maisonneuve.

——. 1994. *Naissance des divinités, naissance de l'agriculture: La révolution des symboles au Néolithique.* Paris: Éditions du Centre National de la Recherche Scientifique.

Chase, A. K. 1989. Domestication and domiculture in northern Australia: A social perspective. In *Foraging and farming: The evolution of plant exploitation,* edited by D. R. Harris and G. C. Hillman, 42–54. London: Unwin Hyman.

Clutton-Brock, Juliet. 1994. The unnatural world: Behavioural aspects of humans and animals in the process of domestication. In *Animals and human society: Changing perspectives,* edited by A. Manning and J. A. Serpell, 23–35. London: Routledge.

Coppinger, Raymond, and Charles K. Smith. 1983. The domestication of evolution. *Environmental Conservation* 10: 283–292.

Corbet, Gordon B., and Juliet Clutton-Brock. 1984. Appendix: Taxonomy and nomenclature. In *Evolution of domesticated animals,* edited by I. L. Mason, 434–438. London: Longman.

Crockford, Susan J. 2000. Dog evolution: A role for thyroid hormone physiology in domestication changes. In *Dogs through time: An archaeological perspective,* edited by S. J. Crockford, 11–20. British Archaeological Reports International Series, 889. Oxford: Archaeopress.

Darwin, Charles. 1868. *The variation of animals and plants under domestication.* Authorized edition. New York: Orange Judd.

Davis, Whitney. 1992. The deconstruction of intentionality in archaeology. *Antiquity* 66: 334–347.

Delaney, Carol L. 1991. *The Seed and the soil: Gender and cosmology in
Turkish village society.* Berkeley: University of California Press.

——. 2001. Cutting the ties that bind: The sacrifice of Abraham and
patriarchal kinship. In *Relative values: Reconfiguring kinship studies,*
edited by S. Franklin and S. McKinnon, 445–467. Durham, NC: Duke
University Press.

Digard, Jean-Pierre. 1990. *L'homme et les animaux domestiques: Anthropologie d'une passion.* Paris: Fayard.

Ducos, Pierre. 1969. Methodology and results of the study of the earliest
domesticated animals in the Near East (Palestine). In *The domestication and exploitation of plants and animals,* edited by P. J. Ucko and
G. W. Dimbleby, 265–275. London: Duckworth.

——. 1978. "Domestication" defined and methodological approaches to
its recognition in faunal assemblages. In *Approaches to faunal analysis
in the Middle East,* edited by R. H. Meadow and M. A. Zeder, 53–56.
Cambridge, MA: Peabody Museum, Harvard University.

Fiddes, Nick. 1991. *Meat: A natural symbol.* London: Routledge.

Flannery, Kent V. 1969. Origins and ecological effects of early domestication in Iran and the Near East. In *The domestication and
exploitation of plants and animals,* edited by P. J. Ucko and G. W.
Dimbleby, 73–100. London: Duckworth.

Franklin, Sarah. 2001. Biologization revisited: Kinship theory in the
context of the new biologies. In *Relative values: Reconfiguring kinship
studies,* edited by S. Franklin and S. McKinnon, 302–325. Durham,
NC: Duke University Press.

——. 2003. Re-thinking nature-culture: Anthropology and the new
genetics. *Anthropological Theory* 3 (1): 65–86.

Franklin, Sarah, and Susan McKinnon. 2001a. Relative values: Reconfiguring kinship studies. In *Relative values: Reconfiguring kinship studies,*
edited by S. Franklin and S. McKinnon, 1–25. Durham, NC: Duke
University Press.

——, eds. 2001b. *Relative values: Reconfiguring kinship studies.* Durham,
NC: Duke University Press.

Galton, Francis. 1865. The first steps towards the domestication of
animals. *Transactions of the Ethnological Society of London* 3: 122–138.

Gautier, Achilles. 1993. "What's in a name?" A short history of the
Latin and other labels proposed for domestic animals. In *Skeletons in
her cupboard: Festschrift for Juliet Clutton-Brock,* edited by A. T. Clason,
S. Payne, and H.-P. Uerpmann, 91–98. Oxford: Oxbow.

——. 1997. Once more: The names of domestic animals. *Anthropozoologica* 25–26: 113–118.

Gentry, Alan W., Juliet Clutton-Brock, and Colin P. Groves. 1996. Proposed conservation of usage of 15 mammal specific names based on wild species which are antedated by or contemporary with those based on domestic animals. *Bulletin of Zoological Nomenclature* 53 (1): 28–37.

Geoffroy Saint-Hilaire, Isidore. 1861. *Acclimatation et domestication des animaux utiles.* 4th edition. Paris: Librairie Agricole de la Maison Rustique.

Goody, Jack. 1973. Bridewealth and dowry in Africa and Eurasia. In *Bridewealth and dowry,* edited by J. Goody and S. J. Tambiah, 1–58. Cambridge: Cambridge University Press.

——. 1977. *The Domestication of the savage mind.* Cambridge: Cambridge University Press.

Groves, Colin P. 1995. On the nomenclature of domestic animals. *Bulletin of Zoological Nomenclature* 52 (2): 137–141.

Hahn, Eduard. 1896. *Die Haustiere und ihre Beziehungen zur Wirtschaft des Menschen.* Leipzig: Duncker and Humblot.

Halstead, Paul L. J. 1996. The development of agriculture and pastoralism in Greece: When, how, who and what? In *The origins and spread of agriculture and pastoralism in Eurasia,* edited by D. R. Harris, 296–309. London: University College London Press.

Haraway, Donna J. 1997. *Modest_Witness@Second_Millennium. FemaleMan©_Meets_OncoMouse™: Feminism and Technoscience.* New York: Routledge.

Harlan, Jack. 1967. A wild wheat harvest in Turkey. *Archaeology* 20: 197–201.

Harlan, Jack R. 1995. *The living fields: Our agricultural heritage.* Cambridge: Cambridge University Press.

Haudricourt, André G. 1962. Domestication des animaux, culture des plantes et traitement d'autrui. *Homme* 2 (1): 40–50.

Hecker, Howard M. 1982. Domestication revisited: Its implications for faunal analysis. *Journal of Field Archaeology* 9 (2): 217–236.

Hillman, Gordon C., and M. Stuart Davies. 1999. Domestication rate in wild wheats and barley under primitive cultivation: Preliminary results and archaeological implications of field measurements of selection coefficient. In *Prehistory of Agriculture: New experimental and ethnographic approaches,* edited by P. C. Anderson, 70–102. Los Angeles: Institute of Archaeology, University of California, Los Angeles.

Hodder, Ian. 1987. Contextual archaeology: An interpretation of Çatal Hüyük and a discussion of the origins of agriculture. *Bulletin of the Institute of Archaeology (London)* 24: 43–56.

———. 1990. *The domestication of Europe: Structure and contingency in Neolithic societies.* Oxford: Basil Blackwell.

Holden, Clare J., and Ruth Mace. 2003. Spread of cattle led to the loss of matrilineal descent in Africa: A coevolutionary analysis. *Proceedings of the Royal Society of London, Series B* 270 (1532): 2425–2433.

Ingold, Tim. 1980. *Hunters, pastoralists, and ranchers: Reindeer economies and their transformations.* Cambridge: Cambridge University Press.

———. 1987. *The appropriation of nature: Essays on human ecology and social relations.* Iowa City: University of Iowa Press.

———. 1994. From trust to domination: An alternative history of human-animal relations. In *Animals and human society: Changing perspectives,* edited by A. Manning and J. A. Serpell, 1–22. London: Routledge.

Jarman, Michael R. 1977. Early animal husbandry. In *The early history of agriculture,* edited by J. Hutchinson, J. G. D. Clark, E. M. Jope, and R. Riley, 85–94. Oxford: Oxford University Press.

Lauwerier, Roel C. G. M., and Ina Plug, eds. 2003. *The Future from the past: Archaeozoology in wildlife conservation and heritage management.* Oxford: Oxbow.

Lawrence, Elizabeth A. 1982. *Rodeo: An anthropologist looks at the wild and the tame.* Knoxville: University of Tennessee Press.

Leach, Edmund R. 1964. Anthropological aspects of language: Animal categories and verbal abuse. In *New directions in the study of language,* edited by E. H. Lenneberg, 23–63. Cambridge, MA: MIT Press.

Leach, Helen M. 2003. Human domestication reconsidered. *Current Anthropology* 44 (3): 349–368.

Lévi-Strauss, Claude. 1966. *The savage mind.* Chicago: University of Chicago Press.

Lyman, R. Lee. 1996. Applied zooarchaeology: The relevance of faunal analysis to wildlife management. *World Archaeology* 28 (1): 110–125.

Mangelsdorf, Paul C. 1958. Ancestor of corn. *Science* 128 (3335): 1313–1320.

Mason, Otis T. 1966[1895]. *The origin of invention: A study of industry among primitive peoples.* Cambridge, MA: MIT Press.

Meadow, Richard H. 1983. Les preuves ostéologiques de la domestication des animaux. *Nouvelles de l'Archéologie* 11: 20–27.

Morales Muñiz, Lola C., and Arturo Morales Muñiz. 1995. The Spanish bullfight: Some historical aspects, traditional interpretations, and comments of archaeozoological interest for the study of the ritual slaughter. In *The symbolic role of animals in archaeology*, edited by K. Ryan and P. J. Crabtree, 91–105. Philadelphia: Museum Applied Science Center for Archaeology, University of Pennsylvania Museum of Archaeology and Anthropology.

O'Connor, Terence P. 1997. Working at relationships: Another look at animal domestication. *Antiquity* 71 (271): 149–156.

Placher, William C. 1996. *The domestication of transcendence: How modern thinking about God went wrong.* Louisville: Westminster John Knox Press.

Price, Edward O. 2002. *Animal domestication and behavior.* New York: CAB International.

Pumpelly, Raphael. 1908. *Explorations in Turkestan, expedition of 1904: Prehistoric civilizations of Anau, origins, growth, and influence of environment*, vol. 1. Washington, DC: Carnegie Institute of Washington.

Reed, Charles A. 1961. Osteological evidences for prehistoric domestication in southwestern Asia. *Zeitschrift für Tierzuchtung und Zuchtungsbiologie* 76: 31–38.

Reitz, Elizabeth J. 2004. "Fishing down the food web": A case study from St. Augustine, Florida, USA. *American Antiquity* 69 (1): 63–83.

Rindos, David. 1984. *The origins of agriculture: An evolutionary perspective.* New York: Academic Press.

Roth, H. Ling. 1887. On the origin of agriculture. *Journal of the Anthropological Institute of Great Britain and Ireland* 16: 102–136.

Russell, Nerissa. 1998. Cattle as wealth in Neolithic Europe: Where's the beef? In *The archaeology of value: Essays on prestige and the processes of valuation*, edited by D. W. Bailey, 42–54. British Archaeological Reports, International Series, 730. Oxford: Archaeopress.

——. 2002. The wild side of animal domestication. *Society and Animals* 10 (3): 285–302.

Russell, Nerissa, and Bleda S. Düring. in press. Worthy is the lamb: A double burial at Neolithic Çatalhöyük. *Paléorient.*

Serpell, James A. 1986. *In the company of animals: A study of human-animal relationships.* Oxford: Basil Blackwell.

Shell, Marc. 1986. The family pet. *Representations* 15: 121–153.

Sigaut, François. 1980. Un tableau des produits animaux et deux hypothèses qui en découlent. *Production Pastorale et Société* 7: 20–36.

Strathern, Marilyn. 1992a. *After nature: English kinship in the late twentieth century, The Lewis Henry Morgan lectures.* Cambridge: Cambridge University Press.

——. 1992b. *Reproducing the future: Essays on anthropology, kinship, and the new reproductive technologies.* New York: Routledge.

Tani, Yutaka. 1996. Domestic animal as serf: Ideologies of nature in the Mediterranean and the Middle East. In *Redefining nature: Ecology, culture and domestication,* edited by R. F. Ellen and K. Fukui, 387–415. Oxford: Berg.

Tapper, Richard. 1988. Animality, humanity, morality, society. In *What is an animal?,* edited by T. Ingold, 47–62. London: Unwin Hyman.

Tringham, Ruth E. 1991. Households with faces: The challenge of gender in prehistoric architectural remains. In *Engendering archaeology: Women and prehistory,* edited by J. M. Gero and M. W. Conkey, 93–131. Oxford: Basil Blackwell.

Uerpmann, Hans-Peter. 1993. Proposal for a separate nomenclature of domestic animals. In *Skeletons in her Cupboard: Festschrift for Juliet Clutton-Brock,* edited by A. T. Clason, S. Payne, and H.-P. Uerpmann, 239–241. Oxford: Oxbow.

Webster, Noah. 1913. *Webster's revised unabridged dictionary.* Springfield, MA: G and C. Merriam.

Wilson, Peter J. 1988. *The domestication of the human species.* New Haven, CT: Yale University Press.

Zeuner, Frederick E. 1963. *A history of domesticated animals.* New York: Harper and Row.

Zohary, Daniel, Eitan Tchernov, and Liora R. Kolska Horwitz. 1998. The role of unconscious selection in the domestication of sheep and goats. *Journal of Zoology* 245: 129–135.

Animal Interface: The Generosity of Domestication

Nigel Clark

It hardly needs to be said that this is not a good time to view the fate of animal domestication with great hope or satisfaction. Although companion species may enjoy certain privileges, especially when their human handlers live amidst relative abundance, those animals that are valued for their flesh or other parts tend to be denied such creature comforts. Market-driven pressures to minimize inputs and maximize outputs of animal bodies have led to increasingly industrialized agricultural practices in which technologies of control and modification are applied to ever-more-intimate aspects of biological being. Factory farming, doses of growth hormones and antibiotics, genetic modification: this is the price animal domesticates pay for our savings at the supermarket checkout. As philosopher Jacques Derrida claims, "no one can deny the unprecedented proportions of this subjection of the animal. ... Everyone knows what the production, breeding, transport, and slaughter of these animals has become" (2002: 394).

If such forms of violence or violation are difficult to stomach when this processing of animal bodies proceeds according to plan, there is worse to behold when the system breaks down. With what appears to be growing regularity, diseases are breaking out amongst farmed animals. Foot and mouth disease in the United Kingdom, France, Brazil and much of East Asia; avian influenza in numerous outbreaks over several continents: each new epidemic accompanied by the extermination of the infected or at-risk animal population. Ending in mass burnings and anonymous burials, these events are an unpleasant reminder how far "domestication" has strayed from its association with the cozy hearth and sheltering enclosures of the *domus*.

What are we to make of this dark underbelly of our desire for frequent, affordable flesh? One way of looking at domestication is to see it as a shortening and tightening of nutrient cycles: an imposition of "efficiency" that seeks to exclude links in the food chain that come between human consumers and those living things they wish to consume (De Landa 1997: 1008). Viewed in this way, domestication appears as an anticipation or prototype of the kind of "economic" logic that is a definitive feature of the era we call "modernity." There are many ways of defining what it is to be "modern," but to put it simply we might say that it is a way of thinking and doing that likes to know its goals, and sets out to attain them in the most efficient and speedy manner. To live in this way, as Michel Foucault put it, is to impose a "grid of intelligibility" on the world from which nothing is supposed to escape (1990: 93). It is to apply a calculus to life and labor such that the value of all things can be known and the costs or benefits of any action discerned—preferably in advance.

Over recent decades these contours of modernity have come into sharper focus, although more often from its failures than its successes. Much has been said about the way that the modern quest for order, clarity, and mastery has undone itself; how it has generated new forms of mess, confusion, and insecurity. It is becoming apparent that "economies" that seek to cycle inputs into outputs with greater speed and tighter control often confound their own logic. Theorists of "risk society" such as Ulrich Beck (1992, 1995) and Anthony Giddens (1994, 1999) have noted how industrial modernity's drive to extend its command over the physical world to ever new depths and degrees is revealing unintended consequences—in the form of runaway events. Mishaps such as nuclear meltdowns, chemical spills, and atmospheric carbon build-up are characterized by "creeping, galloping and overlapping despoliation," which confer on our era a new and frightening profile of endangerment (Beck 1995: 109). If such events are indicative that matter and energy has a tendency to escape from the pathways into which we have tried to confine it, so too do the recent misadventures of industrialized animal husbandry provide a reminder that life itself is a force that often refuses to stay in the grooves and grids we have laid out for it.

In the "runaway" or "undelimitable" event depicted by risk theorists, the linear relationship of cause and effect is derailed: something enters the equation that was not accounted for, something leaves the circuit that was unanticipated. But the excess or exorbitance that has come to haunt the modern utilitarian calculus need not always be so tragic and fearful. Some philosophers and social theorists have found much to affirm amidst the goings-on that exceed the realms of calculativity and knowing. They find in the shadow of our projects other ways of being and doing that are not—cannot ever be—fully encompassed by the dominant "economic" logic. And they seek to make something more of these moments.

If breakdowns or outbreaks ensure things do not always go according to plan, so too can breachings, ruptures, or collisions help change us and our world, especially when we look further afield than the kinds of high-tech accidents that garner so much media and critical attention. As theorists have recently reminded us, it is the other people or things we meet "accidentally," the unanticipated events we get caught up in, the pathways that unexpectedly cross our own, which often change our lives most dramatically. It is when we cease to weigh up the consequences or take the usual measures, they add, that we are most likely to let others draw us out of our "selves" and the circles we usually move in. Such transformative openings of one to an other are referred to using a number of different concepts—"giving," "generosity," "hospitality," "care," "affection," "love." So familiar as to seem trite, these are also amongst the most ancient of philosophical themes. Revitalized in contemporary thought, they are returning as inescapable and often welcome dimensions of ethical, political, legal, and economic life.

It is not simply that we are being called on to crash through the closed circuits of knowledge production or capital accumulation with extravagant gestures of altruism. It is as much that we learn to see and acknowledge the relations of giving and taking, caring and being cared for, hosting and visiting that are always already at play in the more official economies we partake in. In particular, the renewed philosophical interest in generosity and generativity draws attention to the embodied nature of these openings between selves and others, to the inevitable "debt" that any body owes the other bodies who come before and beside it. As feminist philosopher Rosalyn Diprose puts it: "insofar as I am a self, the giving of corporeality is already in operation" (2002: 54). Whether it is a matter of picking up new skills, sexual pleasuring, or organ donation, Diprose claims, the possibilities of our own embodiment are realized through our exchanges with other bodies. Or in the words of Thomas Wall, arguing along similar lines, a "*self* is borrowed, eaten, absorbed from others" (1999: 42).

The field of animal domestication, I want to suggest, offers fertile ground for exploring the give and take, the eating and absorption, that links different kinds of bodies. Although Diprose's notion of "corporeal generosity" focuses on the interplay between *human* bodies, other philosophers have broached the issue of interspecies exchanges: raising questions about the responses and responsibilities that might arise out of our engagement with other living beings. In this regard, the cross-disciplinary study of domestication is already rich in traces of "generous" or mutually transformative relations between species. In particular, the willingness of researchers in the field of animal domestication to consider at once behavioral and somatic shifts, intended and unintentional transformations, and changes in humans as well as other animals promises much for a conversation with philosophers of "corporeal generosity."

As Diprose argues, what a body "owes" to other bodies defies any final reckoning or settling of accounts (2002: 4). Any opening between bodies is always, to some degree, unpredictable, which can make it next to impossible to separate out a risk from a potentiality. This means, as Derrida notes, that it may be difficult to distinguish giving from taking, or to tell a guest from a "parasite" (2000: 59). Where the encounter is between different species, as we will see, such undecidables or uncertainties may be at their extreme. My excursion into the interspecies give and take of animal domestication, then, touches on generosity and affection as sources of creative and life-giving transformations. But it also delves into the realm of poisoned gifts: exchanges that may be damaging or deadly—including those that are "parasitic" in the most literal sense. To move beyond "restricted economies," I propose, is to give other forms of relationality their due—although it is not to imagine that the interface between species could ever fully escape violence or violation.

Unsettling the Colonial Periphery

"The gift," Derrida writes, "should overrun the border, to be sure, toward the measureless and excessive" (1992b: 91). My interest in the "excesses" of animal domestication arises out of an engagement with a rather literal kind of overrunning of borders: namely, the tendency of introduced domesticates to "jump the fence" during the settlement of the colonial periphery. The attempt to transform the biota and landscapes of Europe's "settler colonies" through the introduction of plant and animal species from other regions was a vital and momentous aspect of the colonization process. The process known as "acclimatization" played an important part in the integration of these regions into an emerging global economy, and more generally it formed a cornerstone of the European project of enlightened and progressive improvement of previously "uncivilized" lands.

Needless to say, this colonial project has been subjected to consider-able scrutiny and critique over the intervening years. One of the ways of challenging the foundational narratives of colonialism, one that I have been drawn into, is to examine how the logic of "enlightened" change displayed a distinct tendency to undercut its own principles or precepts. The fate of plant and animal introductions, and sometimes the very motivation for introducing species from other regions, is one of the more striking ways that actual practices of exploration and colonization contravened the whole idea of laying down a "grid of intelligibility." Aside from the fact that much of the transmission of "alien" or "exotic" life was entirely accidental, the impact of introductions can hardly be said to have contributed to orderliness or intelligibility. Many acts of acclimatization were doomed to failure through their abject inappro-priateness, whereas others were destined for a catastrophically successful proliferation (Clark 2002, 2003).

The disorderliness, indeed contrariness, of the practice of species intro-duction appears especially pronounced in the event of domesticates that are released, or "overrun the border" of their own volition, and turn "feral." Formerly domesticated animals that established viable breeding populations independent of human influence opened themselves to the selective pressures of a novel environment, resulting in both behavioral and morphological changes. Feral pigs in Aotearoa New Zealand and Australia, for example, soon turned coarse haired, sharp snouted, long legged, razor tusked, and disarmingly fierce (Clark 2004). Although an evolutionary biologist or zooarchaeologist might see such changes as the fairly predictable resurgence of repressed phenotypes, for a mainstream social scientist such rapid and visible mutability in the realm of the "natural" world may offer more of provocation.

As a sociologist, the whole dynamic of species introduction—with all its shocks and surprises—was a trigger to questioning some of my "inherited" assumptions about the logic of modernity and modern-ization. It prompted me to be more sensitive toward moments of dis-order, opacity, and unpredictability, and it piqued my interest in forms of "economy" other than the rational or "restricted." More than this, the active part played by animals in their own de-domestication and the transformations that resulted seemed to embody a degree of agency and creativity that was much more often attributed to human sociocultural life, or, indeed, was one of the characteristics that was supposed to distinguish human existence from an "unmotivated" natural world.

In this way, my encounters with the unruliness of introduced animal life on the colonial periphery drew me into the longer and broader history of domestication, with an eye to the constitutive role played by deviations, accidents, and irruptions. And it propelled me in the direction of disciplines and subdisciplines that were comfortable with the idea that human and other-than-human forms of life shared important capacities for generative interrelations with the world around them.

Animals, Affection, and Otherness

In this section, I pursue the idea, in a more general sense, of domestication as an unrestricted or other-than-rational economy. My intention is not so much to discredit the notion of more calculating economies as to help loosen their hold on the Western imagination. A belief in well-ordered and all-inclusive systems, the quest for a common measure to smooth the transactions between different kinds of people and different kinds of things, should not be made light of. These are, after all, the underpinnings of the political model of democracy, the economic principle of just rewards for labor, and the tenet of equality before the law: ideals that many of us probably take for granted and few of us would wish to dispense with (see Derrida 1992a).

It is one thing to appreciate the value of well-computed flows and fair exchanges, however, and another to see the rule of utility and necessity as all pervading. The problem with viewing human interaction in this way, critics have noted, is not only that it has a tendency to subsume all the activities and processes present in the modern world into a singular logic but also that it projects itself onto social worlds outside the sphere of Western modernity. In this way, every conceivable social transaction begins to look like a precursor or proxy of "the spirit of calculation" (Bourdieu 1997: 235; see also Derrida 1992b).

Even more telling is that there has been frequent recourse to the model of a restricted economy to explain the workings of the biophysical world. Nature, too, comes to be construed as a realm of scarcity and limitation, in which resources circulate without excess or remainder, every input is carefully recycled, and nothing is squandered. "In the mirror of the economic," writes Jean Baudrillard, "Nature looks at us with the eyes of necessity" (1975: 58). Again, we should be wary of too thoroughly dismissing this idea, and reerecting a boundary between the social and the natural. As ecologists or physicists would contend, certain kinds of circulation and reuse *are* a constant in physical systems. But this need not mean that all expenditure is useful or productive (Bataille 1991: 27–33). And neither does the fact that certain practices or processes appear to have uses or functions necessarily explain how or why they first emerged, as Friedrich Nietzsche pointed out long ago (1956/1870–71: 209–211).

The study of animal and plant domestication is one amongst many fields in which questions of utility and need have been prominent. This is hardly surprising, given the extent to which famine and malnutrition have stalked our species, and the undeniable evidence that agriculture can support many more people per square mile than hunting and gathering. Although it may be some time since scholars have espoused theories of the ascendance of a calculating and motivated "rational man" (see Ingold 2000: 27, 63), we might still discern the traces of the restricted economic imagination in the theorization of domesticatory practices. Both Sandor Bökönyi and Juliet Clutton-Brock's oft-cited definitions of animal domestication, for example, foreground human control or mastery of animals in the interests of profit (see Clutton-Brock 1994: 26; Russell 2002: 287). Further, the debate between theorists who stress human domination, and those who emphasize the symbiotic or commensal dimensions of domestication, seems to hinge on whether the benefits and costs of the relationships in question are equally or asymmetrically distributed (O'Connor 1997; cf. Clutton-Brock 1994).

Alongside the question of how, and for whom domestication "works," however, there is a growing willingness to acknowledge that significant moments in the wider field of domestication exceed such evaluations. Terry O'Connor, in this regard, affirmatively cites Michael Ryder's claim that domestication probably emerged from predator–prey relationships: "almost as an ecological accident. It was almost certainly not conscious or purposeful" (O'Connor 1997: 152). In a related sense, Stephen Budiansky notes that the human side of the domestication process is contingent on biological availability: "humans may select, but only from a set of options determined by forces beyond their control" (1999: 50). Offering the timely reminder that evolution, as Darwin outlined it, has no goal or plan, he further cautions that the motives behind relationships of coevolution are often opaque or ambivalent (1999: 28, 58).

An alternative to necessity is likewise offered by Temple Grandin and Mark Deesing as they speculate about an originary moment in the taming of the wolf—an episode involving an encounter between a hunter and a litter of pups:

> The pups are all frightened and huddle close together as he kneels in front of the den … all except one. The darkest pup shows no fear of the man's approach. … After a mutual bout of petting by the man and licking by the wolf, the man suddenly has an idea. (1998: 1)

Pondering the affective or nonutilitarian beginnings of the process of domestication in this way is by no means new. Writing in the 1860s, Francis Galton proposed that animal domesticates were initially raised in the caring and protective manner that we now associate with pet keeping (Anderson 1998: 122). This view has recently been revisited by James Serpell. After considering some objections to Galton's hypothesis, Serpell affirms that it is "likely that all our currently domestic species, as well as many which were never domesticated, began their association with humans in this essentially non-economic role" (1989: 18–19). Speaking more generally about a disposition toward animals that he believes can be found wherever humans associate closely with other species, Tim Ingold extends this sense of an affective relationship. "We might speak of a history of human *concern* with animals," he writes, "in so far as this notion conveys a caring, attentive regard, a 'being with'" (2000: 76).

The acknowledgement of the importance of emotive ties between humans and animal domesticates takes a leap forward in Donna Haraway's *Companion Species Manifesto* (2003). Exploring the contemporary and historical relationship between "canid" and "hominid," Haraway argues that the disciplinary imperative of training can be the basis of a close emotional bond between dogs and their human handlers (2003: 61–62). But the broader point she makes about the communion between different species extends beyond the dog–human attachment. The "otherness" of species not our own, in this sense, is not taken to be a barrier to affective relations, but is seen as a foundation for ethical and emotional relating. Haraway delves into "the deep pleasure, even joy, of sharing life with a different being," and she has no hesitation about referring to the relationship between humans and their companion animals as one of love (2003: 37).

Considering the potential functions of pet keeping, Serpell observes that, "like any activity, the net benefits must be weighed against the costs" (1989: 17). Although Haraway also points to the advantages and drawbacks of the association with "significant other" animals, it is precisely her move away from the necessity of such accounting that seems most provocative. As she would have it, to enter into a close relationship with another species is to open a network of unknowable and immeasurable outcomes. The human–dog communion, Haraway suggests, is paradigmatic of the "restless exuberance" of zoological encounters. It offers a case study of "multi-directional flows of bodies and value" in a contingent history that includes play as well as labor, waste alongside loss and gain (2003: 9, 12).

Haraway's notion of an interspecies affiliation evokes "corporeal generosity" in the sense that each participant allows him- or herself to be drawn into an open-ended circuit of affect and transformation. Although it is concerned with a different sort of relationship between species, this sense of multidirectionality and lack of closure has important continuities with what I was saying earlier about the unruly generativity of animal life on the colonial periphery. It also resonates profoundly with the claims Diprose makes about giving. "If the gift opens possibilities for existence," she writes, "then its operation rests on not determining anything about who gives what to whom ahead of or during an encounter" (2002: 55). It is not simply that already constituted "others" have the option of giving to one another, both Haraway and Diprose suggest, but that our very identities as individuated or discernibly different beings arise out of exchanges with those who differ from us. Like it or not, every body relies on the generosity of other bodies, not only in the sense of what is corporeally bequeathed by parents and forebears, but also through that which is taken on by processes of imitation, or incorporated through the material transactions we have with others. We are what we ingest, absorb, or appropriate, in other words.

The Animal Interface

A social life that encompasses domesticated animals, in this light, can be seen to rest more primordially on a kind of mutual *dis*possession than on the possession of animals by human actors; a letting go of customary precautions and boundary maintenance on the part of each participating species. Whatever benefits and utilities might eventually emerge, any ongoing interspecies association, it might be argued, hinges on "a gift of the possibility of a common world" (Diprose 2002: 141). This brings to mind the insights of the ethical philosopher Emmanuel Levinas on the primacy of a "non-allergic" reaction—a response that renounces violence or hostility—in the forging of a relationship with the other (1969: 199). Whereas Levinas reserved his concerns for the interhuman realm, his considerations on the significance of the "caress," as opposed to the act of grasping or seizing, invite extension to the human encounter with animal others. The caress, he suggests, "does not know what it seeks," it expresses a desire provoked by otherness that lacks clear purpose or plan (Levinas 1987: 89).

John Llewelyn (1991) takes this "other-than-human" reading of Levinas further—building on the significance of the face in the French philosopher's writings. For Levinas, the face—and in particular, the eyes—is the most immediate way we perceive vulnerability and need in the other: the face thus standing for the frailty inherent in embodied existence (Llewelyn 1991: 63). Although he concedes that animals suffer, and that the ethical should extend to all living beings, Levinas is nevertheless circumspect. The significance of the suffering we witness on the face of the other invites a conversation: an opening Levinas is reluctant to extend beyond the potential for dialogue of our own species: "The human face is completely different and only afterwards do we discover the face of an animal" (Llewelyn 1991: 65). On this count, Llewelyn is not convinced. Downplaying Levinas's stress on speech, he argues that the face of an animal can appeal to us, even call us into question. And in this light he takes inspiration from the poet Rilke, who went so far as to claim that "What *is* outside, we know from the animal's face alone" (Llewelyn 1991: 157). Taking his "reanimation" of Levinas in the direction of political ecology, Llewlelyn concludes by pointing to our responsibility to preserve the conditions for the flourishing of other beings—an obligation that for him encompasses both domesticated and free-ranging creatures (1991: 254–255).

It is one thing, however, to foreground human responsiveness to appealing and vulnerable fellow creatures. It is quite another to speculate about corresponding sensibilities moving in the other direction. Osbjorn Pearson gives us serious grounds for doubt. "Humans are remarkable for the amount of co-operation, sharing, and reciprocal altruism that typifies our societies," Pearson asserts, "a similar trend toward co-operation and sharing with conspecifics does not characterize domestic animals" (Leach 2003: 362). The philosopher Martin Heidegger, whose writings Levinas frequently engaged with, was equally forthright. For him, what distinguishes animals from humanity is not only their lack of a capacity for thought and language, it is also their inability to bestow gifts (Wyschogrod 1990: 82).

But a lot rests on how we choose to define "giving." The sense in which Diprose and others constitute the gift—as a kind of excessive and often nonvolitional flow between bodies—suggests a more inclusive reading of the offerings of other forms of life. Similarly playing on the notion of our susceptibility to be moved by otherness, Alphonso Lingis proposes that we humans acquire many of our gestures, postures, and desires through communion with animals. And it is not simply, or even primarily, our glimpses of charismatic free-living fauna that has such affect, so much as the ongoing and intimate exchanges we have with more familiar creatures. Lingis has us developing our sensual and emotional registers through such experiences as our infantile fondling of kittens or lambs, our childhood observations of hens defending their chicks, our memories of mounting the "smooth warm flanks of a horse" (2000: 36–37).

Such practices of mimesis and projection offer one way of approaching the question of how "the organization of the body (is) given to and by the corporeality of others" in the context of codwelling species (Diprose 2002: 69). Along with Llewelyn's notion of the ethical appeal in the face of the animal, however, Lingis's unashamed eroticizing of our encounters with other creatures may risk alienating those with a more scientific approach to the phenomenon of domestication. But there are alternative modes of engagement with the interface of humans and associate species that seem to substantiate the idea of the mutual affectivity of neighboring bodies—inquiries that draw more on archaeological evidence than discourses of ethical philosophy. It is broadly agreed amongst scholars in anthropology and animal science that there is a set of characteristics found in domesticated mammals that helps distinguish them from free-ranging counterparts. Such changes include loss of skeletal robustness, shortening of the muzzle or facial region, and retention of juvenile behaviors into adulthood (Leach 2003: 349). What has frequently been passed over in this context, as Helen Leach (2003) has recently noted, is that animal domesticates are not alone in such bodily modifications.

Corporeal Generosities

"We turn into our partners, and even our dogs, just by dwelling with them," Diprose observes (2002: 70). Such mutual influence, however, may go well beyond mere traffic in gesture and expression. Leach directs our attention to the parallel between the somatic transformations observed in the archaeozoological record of animals going through the early phases of domestication, and changes noted in human morphology over corresponding periods—pointing to the shared shift from robustness to gracility that is especially evident in the face and head. A key factor in this convergence, she suggests, is the cultural modification of the environment in ways that protect both humans and their livestock from many of the physical challenges—and thus the selective pressures—associated with a more free-ranging existence. "For the human, the combination of adoption of a built environment, change in diet consistency, and lowered mobility brought about morphological changes similar to those seen in domestic animals" (Leach 2003: 360).

Leach's thesis does not rule out other explanations for changes in bodily form of domesticates that imply greater human intentionality, such as the selection of smaller, more easily handled animals (Leach 2003: 350). What it does do, though, is to grapple with the thorny and frequently bypassed issue of humankind unintentionally "domesticating itself" along with its animal associates. From the perspective of an "economy" of corporeal generosity this raises the prospect that by "giving" shelter and protection to other animals humans precipitated bodily transformations shared with these other species, changes that could not have been intended or anticipated.

But there is an even more provocative sense in which we might draw a connection between convergent human–domesticate evolution and the generous, receptive attitude to others affirmed in the work of Levinas, Lingis, Diprose, and fellow ethical philosophers. As Leach points out, a number of the transformations she investigates—including "craniofacial" reduction and general loss of robustness—have been linked to the phenomenon of neotony or paedomorphosis—which entails the retention into adulthood of certain features associated with juvenility (2003: 354). Evidence suggests that these morphological changes are related to nonaggressive and tolerant behavior, which are likewise characteristic of immature animals. This conclusion is supported by Belyaev and Trut's account of a silver fox domestication project that aimed for docility and tolerance of humans but unexpectedly produced accompanying craniofacial reductions (Leach 2003: 354–355).

Although the connection between neotony and domestication remains uncertain, and indeed contentious (see Price 1998: 49), it raises intriguing possibilities for drawing together the question of animals "lending" themselves to domestication and the issue of the openness of humans to closer bonds with other species. For as Konrad Lorenz proposes, humans, too, display some familiar neotonic characteristics:

> I am convinced that man owes the life-long persistence of his constitutive curiosity and explorative playfulness to a partial neoteny which is indubitably a consequence of domestication. (Grandin and Deesing 1998: 20)

Stephen Budiansky (1999) agrees that significant features of neotony may be present in humans as well as their domesticates. But Budiansky makes the claim that the behavioral and somatic changes that characterize neotony are not simply an outcome of domestication but an important prerequisite. He argues that, prior to their more structured association, humans and their livestock spent many millennia familiarizing themselves with each other, driven together by the climatic upheavals of the Pleistocene era (1999: ch. 4). Fluctuating environmental conditions, Budiansky reminds us, favor life-forms that can vacate and colonize ecological niches rapidly: which is to say species that are adaptable and opportunistic (1999: 73–75). Such conditions, he notes, favor the evolutionary strategy of neotony, which brings with it the curiosity and rapid learning ability of the young, including "a non-discriminating willingness to associate and play with members of other species" (Budiansky 1999: 78). The retention into adulthood of these juvenile traits, Budiansky suggests, offers a platform for the emergence of the relationship of domestication. And this applies to humans no less that their potential domesticates: "The neotony that is part of our own evolutionary heritage may have likewise made us more willing to enter into relationships with animals other than the highly specialized one of predator to prey" (Budiansky 1999: 80).

These observations might cast some light on incidents of interspecies association that were often witnessed on the colonial periphery. Farmer and naturalist Herbert Guthrie-Smith, a keen-eyed observer of rural New Zealand under colonial transformation, recounts the first arrival of an introduced species to his part of the country.

> The attraction of the stag to the spot chosen was doubtless the small herd of wild horses strayed from native villages deserted and never afterwards repeopled. With them the lonely deer formed one of those curious animal friendships that strayed creatures make, a companionship similar to that of another stag which, at a much later date, consorted with the Black Head stud bulls, or to that of the first rabbit seen north of Petane, which for several seasons accompanied a flock of wild turkeys on the Tangoio run. (1999/1921: 337)

These expressions of tolerance or bonding also occurred under conditions of environmental fluctuation and turbulence—only this time the stress was induced by human activities. Having observed the behavior of animals, both domestic and free ranging, as they confronted unfamiliar objects or conditions, Guthrie-Smith concluded: "Curiosity is by no means confined to humanity" (1999/1921: 304).

After a more systematic inquiry, Grandin and Deesing make the claim that there is genetically based natural variation in many free-living animals with regard to responses to novel experience, such as encountering humans—with a minority displaying "a quiet exploratory reaction without either fear or aggression" (1998: 2). This is the scenario illustrated in their hypothetical hunter-meets-wolf cub tale, although it might be added that canid and hominid might best be seen as selecting each other—given that the individuals of *both* species seem endowed with exceptional fearlessness and inquisitiveness. Taken as a precondition of domestication, then, it is not so much the genetic predisposition for placidity and homeliness that appear pivotal, but an openness and receptivity to "otherness."

In this way, insights from the scientific study of evolution resonate with the "non-allergic" response to the other privileged in the writings of ethical philosophers like Levinas, Derrida, and Diprose—fleshing out the notion that a renunciation of hostility is the "gift" from which the possibility of a shared world arises. And although it undoubtedly rests on a too-literal reading of Levinas, it is tempting also to reflect on the primacy of the face, and the resultant accentuation of the eyes, in the morphological shifting linked to neotenic nonaggression. But what we might say, with more confidence, is that the give and take between heterogeneous species exceeds any sense of deliberation or planning: contributing to bodily and behavioral changes with a utility that can only ever be grasped in retrospect. Or as Haraway puts it in regard to the generative interchange between dogs and humans: "Flexibility and opportunism are the name of the game for both species, who shape each other throughout the still ongoing story of co-evolution" (2003: 29).

If taken literally, the idea of a generous and generative "animal interface"—for all that it may implicate differentiated species—would seem to imply at least a modicum of shared physiological, neurological, and limbic faculties. Although a meaningful encounter between living beings need not necessarily involve volition or judgment, mutual recognition calls for a capacity to "read" the other, to register and respond to each other's presence. In this regard, we should not "focus" too strongly on sight or the eyes, recalling the extent to which some animals depend on acuteness of hearing or scent discrimination and perception of movement (Patton 2003: 97). And neither should we prioritize "language," at least in any sense that privileges the human experience of this faculty, especially because the linguistic turn in philosophy has reminded us that even amongst members of our own species, mistranslation and multiple interpretation is rife (see Derrida 2003).

But even the requirement of a minimum of shared sensory faculties as the condition of "intimacy" between members of different species comes into question, if we pursue the unintended consequences of domestication along some of its more minute and surreptitious pathways.

Poisoned Gifts: Domestication and Pathogen Exchange

I have been suggesting that a certain strand in Western philosophy concerned with excessiveness and its expression in the act of giving might be brought into convergence with the archaeological and biological inquiry into the emergence of animal domestication. But there is a sense in which "gifts" and "generosity" are loaded terms, importing an everyday connotation of beneficence that does not always sit comfortably in this context. This is especially so when we consider the dangers faced by any organismic body that permits itself such intimacy with other bodies that mutual influence in behavior or morphology becomes a possibility. "Corporeal generosity," as Diprose puts it, "is writing in blood that says this body carries a trace of the other" (2002: 195). And in this way she reminds us that giving is always risky, that the offering or receiving of a gift, by virtue of the potentiality it conveys, is inevitably a kind of rupture or violation. Or as Derrida proclaims: "Such violence may be considered the very condition of the gift, its constitutive impurity" (1992b: 147).

The word "gift," Marcel Mauss noted, shares the meaning of poison in the Germanic languages, a reminder that the favored present for the ancient Germans was alcoholic (1997: 30). In the annals of close human–animal association, however, it is not poisons but pathogens that manifest the dark underside of the generous encounter. As William McNeill points out: "Most and probably all of the distinctive infectious diseases of civilization transferred to human populations from animal herds" (1998: 69). These "deadly gifts from our animal friends"—otherwise known as zoonoses—include worms, protozoa, bacteria, fungi, and viruses (Diamond 1998: 207). Arguably, their traffic amongst and between species is no less intrinsic to the domestication process than is artificial selection or incidental morphological change.

Pathogens, we might say, play on the terrain of the exorbitant: they are the gift that keeps on giving. Where there is intimacy, there will be microscopic life to-ing and fro-ing between partners, and where the parties themselves happen not to be conspecifics there is an opportunity for microorganisms to move permanently across species boundaries (see Garrett 1995: 572–579). But what might be impartial and dispassionate survivalism at one level may well blossom out of warmth and tenderness at another. It is human "proximity to the animals we love," especially "cuddly species (like young lambs) with which we have much physical contact," Jared Diamond notes, which heighten our risk of pathogen transfer (1998: 207, 213).

In a more general sense, it is the agglomeration of human and animal populations that provides the conditions for contagious diseases to take hold. Such social animals as cows, sheep, and pigs would already have been reservoirs of pathogens prior to domestication; settled agriculture providing the density of hosts—both human and domesticate—to evolve and sustain diseases (Diamond 1998: 205–206). By the same logic, this environment offers rich opportunities for pathogens to jump between species. What Diamond says of measles, a virus likely to have come from cattle, might equally apply to other domestic or companion animals: "that transfer is not at all surprising, considering that many peasant farmers live and sleep close to cows and their feces, urine, breath, sores, and blood" (1998: 206–207). And the movement of these infections, as we might expect, is multilateral: from humans to livestock, amongst different domesticated species, and frequently overrunning the border between domesticates and their free-living relatives (McNeill 1998: 71).

Across Eurasia, human populations gradually came to terms with the diseases they had exchanged with their livestock, an accommodation achieved through the costly selective pressure of successive plagues over thousands of years. The settling into endemicism of these infectious diseases was the prelude to their devastating introduction to the "epidemiologically naive" populations of lands previously insulated by oceans (Crosby 1986: ch. 9; McNeill 1998: ch. 5). It has been estimated that 95 percent of the indigenous population of the Americas perished in epidemics over the century or two following contact with Europeans, a scenario repeated to greater or lesser degree in all other of Europe's "new worlds" (Diamond 1998: 211). As Diamond would have it, the European conquest of these lands and subsequent demographic takeovers "might not have happened without Europe's sinister gift to other continents—the germs evolving from Eurasians' long intimacy with domestic animals" (1998: 214; see also McNeill 1998: 235).

As a descendent of European settlers in a southern hemisphere colony, this catastrophic history has a particular poignancy for me. It is also drives home the point that the enlightening project of modernity not only "overran" its own logic and principles as it spread across the earth's surface, its very extension was usually premised on its shadowy underside of excess and disorder. And as I suggested at the outset of this chapter, the age of poisoned gifts is far from over. Recent years have seen the emergence of a new strain of Creutzfeldt–Jakob disease—linked with Bovine Spongiform Encaphelopathy or "mad cow disease," the frightening but short-lived SARS epidemic, and numerous irruptions of avian influenza (see Davis 2005).

Yet, for all their terrible toll, we might also acknowledge a kind of "generosity" in the way that pathogens take advantage of the proximity and porosity of larger bodies. The "trace of the other" that Diprose sees as constitutive of all bodily identity is nowhere more literally inscribed that in the bequest of successive microorganismic invasions. Invading viruses have left their mark throughout the living world, with hundreds of retroviruses becoming integrated in the human genome, many of which now perform vital defense functions against subsequent infection (Lederberg 2004: 55). Indeed, evolution—our own as much as that of any other species—is partially propelled by infectious microorganisms. As Haraway reminds us:

> Evolutionary biology posits that we only evolve with our illnesses, and
> that the difference and diversity that comes from infection and contagion
> is what actually allows us to continue to proliferate and survive in a
> variety of environmental conditions on the planet. (2000: 22)

Just as a generous or hospitable relationship between animal others is premised on a withholding of violence, so too does a lasting host–pathogen association depend on the way bacteria, in the words of biologist Joshua Lederberg "*withhold* their virulence" (2004: 55). For a pathogen to survive, it must avoid too rapidly destroying its host, and in this regard it is in the interests of both species to evolve toward mutual tolerance. "This is what disease as experienced by humans is all about," Lederberg claims "the establishment of a foothold so the obliging host will provide warm food and shelter and be domesticated to the service of that parasite" (2004: 54). Whether we take this metaphorically or literally as a domesticatory relationship, it is a cogent reminder that the larger organisms who enter into the association we more typically call "domestication" are always already the outcome of "generous" encounters: exchanges at once generative and deadly.

Moreover, the rich mutual responsiveness of microbial life and larger organisms suggests that there are ways of communicating that do not depend on anything remotely approximating a common sensory system. The ability of a virus—barely even complex or sensate enough to qualify as "living"—to "read" a host's bodily makeup well enough to confound its immunological system and to appropriate its mechanisms of cellular reproduction—together with the host's ability to develop novel immuno-logical defenses to an uninvited microscopic visitor—hints at just how multilayered "recognition" can be. And at the "closeness" of the relationships that result from this furtive but incessant give and take. In the words of digital media artist Melinda Rackham, herself a Hepatitis C carrier: "a virus penetrating your core is probably the most intimate relation you can have with another species" (2000: 22).

Conclusion

The idea of an embodied generosity hinging on the susceptibility of living beings to the "affect" of other bodies helps turn our attention to the open-endedness of interspecies relations. It reminds us that the adaptability and creativity of living things is not simply an attribute of life in the "wild," and neither is it a capacity that has been entirely appropriated and overwritten by human technological practices. Rather, it is an ongoing process that is found wherever species come into sustained and intimate relationships, whether these are intentional or incidental.

In our coexistence with the diseases that pass within and between species, virologist Stephen Morse warns us to "begin by expecting the unexpected (1994: 325), while philosopher Jacques Derrida, addressing our contemporary condition more generally, advises us to "open the calculation to the uncalculable" (2001: 259). Neither is telling us not to predict, deliberate, or otherwise weigh up our own interests, but both seem to be cautioning about the limits of this kind of "economy," the limits of knowledge and mastery. The idea that our best-laid plans for controlling the biophysical world have unpredictable and incalculable consequences is now a central concern in social theory, and as I noted earlier, theorists like Beck and Giddens see this self-fabricating of risk as one of the definitive features of the current phase of our modernity. "Manufactured risk," as Giddens puts it, "refers to risk situations which we have very little historical experience of confronting" (1999: 26).

Or have we? My allusion to the "creeping, galloping and overlapping despoliation" that so often followed from species introductions on the colonial periphery, and my more general discussion of the unforeseeable consequences of animal domestication, were intended to show that altering the pattern of our associations with different forms of biological life has always been risky. Wherever bodies come close enough to be of benefit to other bodies, I have been suggesting, there will inevitably be a danger of other transmissions and transformations that are threatening or deleterious—for there can be no opening of one living being to another that is entirely predictable. And in this regard, our era is indeed characterized by the threat of the kind of unanticipatable and undelimitable accidents that Giddens, Beck, and others have described. But so too have many other times and places given rise to similar scenarios of risk.

In this sense, a history or archaeology of domestication that is attuned to the inherent excessiveness of interspecies association at once meshes with and perturbs some of the central concerns of contemporary social theory. The expansive temporal and spatial scales that feature in studies of domestication, and the relatively rich tradition of merging social and biophysical variables, I would argue, helps put the more nascent social theoretic concern with the dangers of manipulating life and matter into a much broader context. An important part of this wider contextualization is that it gives us time—historical time, evolutionary time—to register and account for the potentialities that also inhere in these fraught encounters. In this regard, then, the concept of an embodied generosity seems to offer a way of holding open, at once, danger and possibility, the threat of destruction and the chance of generativity. If the animal life we depend on, not to mention our own "animal lives," is the outcome of such generosities, then, along with our fears, we have a lot to be grateful for. How to express that gratitude while continuing to satisfy our appetites remains a challenge with an open and endless horizon.

References

Anderson, Kay. 1998. Animal domestication in geographic perspective. *Society and Animals* 6 (2): 119–134.

Bataille, Georges. 1991. *The accursed share,* vol. 1. New York: Zone Books.

Baudrillard, Jean. 1975. *The mirror of production.* St Louis: Telos Press.

Beck, Ulrich. 1992. *Risk society: Towards a new modernity.* London: Sage.

——. 1995. *Ecological politics in an age of risk.* Cambridge: Polity Press.

Bourdieu, Pierre. 1997. Marginalia—some additional notes on the gift. In *The logic of the gift: Toward an ethic of generosity,* edited by A. D. Schrift, 231–243. New York: Routledge.

Budiansky, Stephen. 1999. *The Covenant of the wild: Why animals chose domestication.* New Haven, CT: Yale University Press.

Clark, Nigel. 2002. The Demon-seed: Bioinvasion as the unsettling of environmental cosmopolitanism. *Theory Culture and Society* 19 (1–2): 101–126.

——. 2003. Feral ecologies: Performing life on the colonial periphery. *Nature performed: Environment, culture and performance,* edited by B. Szerszynski, W. Heim and C Waterton, 163–182. Oxford: Blackwell.

——. 2004. Pigs. In *Patterned ground: Entanglements of nature and culture,* edited by S. Harrison, S. Pile, and N. Thrift, 218–219. London: Reaktion.

Clutton-Brock, Juliet. 1994. The unnatural world: Behavioural aspects of humans and animals in the process of domestication. In *Animals and human society: Changing perspectives,* edited by A. Manning and J. Serpell, 23–35. New York: Routledge.

Crosby, Alfred W. 1986. *Ecological imperialism: The biological expansion of Europe 900–1900.* Cambridge: Cambridge University Press.

Davis, Mike. 2005. *The monster at our door: The global threat of avian flu.* New York: New Press.

De Landa, Manuel. 1997. *A thousand years of nonlinear history.* New York: Swerve Editions.

Derrida, Jacques. 1992a. Force of law: The "mystical foundation of authority." In *Deconstruction and the possibility of justice,* edited by D. Cornell, M. Rosenfeld, and D. G. Carlson, 3–67. New York: Routledge

——. 1992b. *Given time: 1. Counterfeit money.* Chicago: University of Chicago Press.

——. 2000. *Of hospitality.* Stanford. Stanford University Press.

——. 2001. A Roundtable discussion with Jacques Derrida. In *Derrida Downunder,* edited by L. Simmons and H. Worth, 249–263. Palmerston North, New Zealand: Dunmore Press.

——. 2002. The animal that I am (More to follow). *Critical Inquiry* 28 (Winter): 369–418.

——. 2003. And say the animal responded? In *Zoontologies: The question of the animal,* edited by C. Wolfe, 121–146. Minneapolis: University of Minnesota Press.

Diamond, Jared. 1998. *Guns, germs and steel.* London. Vintage.

Diprose, Rosalyn. 2002. *Corporeal generosity: On giving with Nietzsche, Merleau-Ponty, and Levinas.* Albany: State University of New York Press.

Foucault, Michel. 1990. *The history of sexuality,* vol. 1, *An Introduction.* London. Penguin.

Garrett, Laurie. 1995. *The coming plague.* New York: Penguin.

Giddens, Anthony. 1994. Living in a post-traditional society. In *Reflexive modernization: Politics, tradition and aesthetics in the modern social order,* edited by U. Beck, A. Giddens and S. Lash, 56–109. Cambridge: Polity Press.

——. 1999. *Runaway World*. London: Profile Books.

Grandin, Temple, and Deesing, Mark, J. 1998. Behavioral genetics and animal science. In *Genetics and the behavior of domestic animals,* edited by T. Grandin, 31–65. San Diego: Academic Press.

Guthrie-Smith, Herbert. 1999[1921]. *Tutira: The story of a New Zealand sheep station*. Auckland: Godwit.

Haraway, Donna. 2003. *The companion species manifesto: Dogs, people, and significant otherness*. Chicago: Prickly Paradigm Press.

Ingold, Tim. 2000. *The perception of the environment: Essays on livelihood, dwelling, and skill*. New York: Routledge.

Leach, Helen. M. 2003. Human domestication reconsidered. *Current Anthropology* 44 (3): 349–368.

Lederberg, Joshua 2004. Of men and microbes: Understanding SARS and infectious diseases. *New Perspectives Quarterly* 21 (2): 52–55.

Levinas, Emmanuel. 1969. *Totality and infinity*. Pittsburgh: Duquesne University Press.

——. 1987. *Time and the other*. Pittsburgh: Duquesne University Press.

Lingis, Alphonso. 2000. *Dangerous emotions*. Berkeley: University of California Press.

Llewelyn, John. 1991. *The middle voice of ecological conscience: A Chiasmic reading of responsibility in the neighbourhood of Levinas, Heidegger and others*. Macmillan: Basingstoke.

McNeill, William H. 1998. *Plagues and peoples*. New York: Anchor.

Mauss, Marcel. 1997. Gift, gift. In *The Logic of the gift: Toward an ethic of generosity,* edited by A. D. Schrift, 28–32. New York: Routledge.

Morse, Stephen S. 1994. The viruses of the future? Emerging viruses and evolution. In *The evolutionary biology of viruses,* edited by S. S. Morse, 325–336. New York: Raven Books.

Nietzsche, Friedrich. 1956[1870–71]. *The birth of tragedy and the genealogy of morals*. New York: Anchor Books.

O'Connor, T. P. 1997. Working at relationships: Another look at plant domestication. *Antiquity* 71: 149–156.

Patton, Paul. 2003. Language, power and the training of horses. In *Zoontologies: The question of the animal,* edited by C. Wolfe, 121–146. Minneapolis: University of Minnesota Press.

Price, Edward O. 1998. Behavioral genetics and the process of animal domestication. In *Genetics and the behavior of domestic animals,* edited by T. Grandin, 31–65. San Diego: Academic Press.

Rackham, Melinda. 2000. My viral lover. *Real Time/On Screen* 37 (June–July) 22.

Russell, Nerissa. 2002. The wild side of animal domestication. *Society and Animals* 10 (3): 285–302.

Serpell, James. 1989. Pet-keeping and animal domestication: A re-appraisal. In *The walking larder: Patterns of domestication, pastoralism and predation,* edited by J. Clutton-Brock. 10–20. London: Unwin-Hyman.

Wall, Thomas. 1999. *Radical passivity: Levinas, Blanchot and Agamben.* Albany: State University of New York Press.

Wyschogrod, Edith. 1990. *Saints and postmodernism: Revisioning moral philosophy.* Chicago: University of Chicago Press.

Selection and the Unforeseen Consequences of Domestication

Helen M. Leach

However it is defined, domestication was a process initiated by people who had not the slightest idea that its alliance with agriculture would change the face of their planet almost as drastically as an ice age, lead to nearly as many extinctions as an asteroid impact, revolutionize the lives of all subsequent human generations, and cause a demographic explosion in the elite group of organisms caught up in the process (Leakey and Lewin 1996; Vitousek et al. 1997). Such unforeseen consequences are seldom discussed in the literature of domestication, perhaps because it is not in the nature of the species that started the process to admit that it is not in control. This species, *Homo sapiens sapiens,* has recently been reluctant even to contemplate that it might have become domesticated along the way—civilized, yes, but not biologically domesticated like the animals. On the last occasion that such a notion was influential, the supposed degeneracy associated with domestication (reviewed by Sax 1997: 12–13) was used to justify the elimination of many individuals judged "unfit" and the adoption of eugenic programs to improve the chosen (Tort 1996: 1272). Nevertheless, if the meaning of domestication is under new scrutiny, it is important to look at its effects on all the species that participated.

Most definitions of domestication still portray it as a process that was driven by humans to satisfy human needs, often stating or implying in their choice of terms that humans "created" their domesticates through conscious intervention. For example Steven Strauss's (2003) article on genetic engineering and domestication in *Science* used the *American Heritage Dictionary* definition of 1982:

Domesticate—to train or adapt (an animal or plant) to live in a human environment and be of use to humans.

In this reading of the term, which implies active human agency, animals and plants become commodities and artifacts, to the extent that today their genes can be patented, the ultimate affirmation of property rights. Such a hands-on role does not permit recognition of unintended domestication of humans, only the forms of deliberate modification aimed for in human eugenic experiments (Bajema 1982: 203) or from controlled breeding among slaves (Tani 1996).

The active role of humans in the long history of domestication has been stressed in the debate over genetic engineering (e.g., Royal Society 1998: 5). Here, as a response to critics who fear the impact of what they see as untested technology, transgenic manipulation is presented as a refinement of the selection by "countless generations of farmers" of novel and useful plant types. Although the improved types are described as "naturally formed mutants," the process is nevertheless portrayed as "the earliest and simplest form of genetic modification" (Belzile 2002: 1111). Many involved in the debate begin their reports by stating that this form of selection is the foundation of domestication and stretches back 10,000 years. Channapatna Prakash (2001: 10–11) goes further: "Using gene transfer techniques to develop GM crops thus can be seen as a logical extension of the continuum of devices we have used to amend our crop plants for millennia."

To what extent did Neolithic farmers practice deliberate artificial selection to improve the usefulness of certain plants and animals? Or were forms of "unconscious" selection responsible for the majority of observed changes, operating in an "environment of domestication" that disrupted and changed the habitats, nutrition, and group dynamics of the early domesticates? I have previously argued that the humans who shared that environment of domestication were also affected by the new lifestyle, showing cranial and postcranial gracilization, reduction in size, and dental changes similar to those observed in some of their domestic animals (Leach 2003). Does this mean that humans became "domesticated" like their animals?

There are good heuristic reasons for retaining the term "domestication" for this early coevolution, including the importance of the *domus*, or homestead, in providing a setting for the growing interdependence of certain species, including humans. But the term is applied equally to recent human manipulation of plants and animals, referred to as selective breeding or artificial selection. There is growing recognition that this type of systematic breeding emerged only within the past

300 years. With a few notorious exceptions it has not been applied to humans. If selective breeding for improvement is so recent, what was the nature of the domestication relationship in the millennia between the onset of the Neolithic and the 18th-century C.E., when the first stock improvement programs began to spread. What breeding systems were described by the ancient writers on husbandry?

If the selection mechanisms have changed their character over the past 10,000 years, is that one undivided term "domestication" the best to convey the essence of the changing relationships? I intend to put a case for recognizing four stages of domestication based on the predominant modes of selection:

- the first stage, affecting humans and other commensals, in addition to leading to the appearance of the traditional plant and animal domesticates, was characterized largely by "unconscious" forms of selection to cope with new environments and cultural practices;
- the second stage affected traditional plant and animal domesticates, and involved both unintentional selection pressures and the application of nonspecific breeding principles to maintain favored landraces or types that had originally developed as adaptations to particular regions;
- the third stage saw the formation of improved animal breeds and plant cultivars, initially through combinations of crossbreeding and inbreeding, then increasingly through laboratory-based techniques such as in vitro fertilization, induced mutation and polyploidy;
- the fourth and most recent stage now achieves "improvements" through advanced molecular biological techniques that circumvent genetic barriers; it was applied first to plants, then to animals, but is strongly promoted for its potential medical applications to humans.

Because the third and fourth stages occurred over the past 300 years, and are therefore more accessible in texts on crop improvement and animal breeding, the emphasis here will be on the first two stages. Persisting over ten millennia, and predating any systematic understanding of heredity, these early stages are poorly documented, and what evidence exists requires careful interpretation.

Selection in Domestication

Rather than treating humans as animal members of the biosphere and their impacts on the survival of other organisms as just another form of natural selection, we have inherited Charles Darwin's position that, like animals and plants, humans are subject to natural selection, while at the same time, unlike them, they practice two other forms of selection, "unconscious" and "conscious" (which Darwin [1868 I: 214] also called "methodical" selection). Unconscious selection has become the portmanteau term for pressures that are neither natural selection as it is presumed to operate "in the wild," nor artificial, methodical selection as it is currently applied on the farm or in the laboratory. The term is broad, and confusing in that it is now used in a different sense from that set out by Darwin (1868 I: 214). Darwin regarded unconscious selection as the choice of breeding stock simply to maintain good quality, whereas methodical selection was clearly goal-directed, to a predetermined standard. Unconscious selection is now more often used for the selective pressures brought to bear on animals and plants by placing them in a human modified environment and exposing them to particular systems of husbandry (see Table 1). If the morphological changes that such pressures brought about in animals also occurred in humans sharing those environments, then the current usage cannot comfortably encompass humans. The terminology requires both review and revision.

Table 1. Terms applied to selection pressures in domestication

Reference:	Natural ?		? Artificial
Darwin 1868 1:214		Unintentional or unconscious selection – driven by simple desire to maintain good stock	Methodical or conscious selection – to create some improvement already pictured in breeder's mind
Darlington 1956:133	Natural or unconscious selection in cultivation		Plant breeding by persistent and unremitting selection
Darlington 1963:156	Operational (unconscious) selection in cultivation		Intentional artificial selection
Harlan et al. 1973:313,321	Automatic selection		Deliberate selection

Reference:	Natural ?		? Artificial
Donald and Hamblin 1983:101	Natural selection for adaptation to agriculture	Non-specific selection by man	Conscious or methodical selection
Zohary 1984:579	Automatic, unconscious selection		Deliberate, conscious selection
Hanelt 1986:193	Natural selection for adaptation to agriculture		Artificial, conscious selection
Heiser 1988:77–8	Non-intentional human selection, = unconscious (non sensu Darwin) = automatic = operational		Methodical, conscious artificial selection
Bökönyi 1989:26	Unintentional breeding selection		Developed (animal) breeding with conscious selection for productivity and breed
Tchernov and Horwitz 1991:57	Unconscious, unintentional selection		Conscious, deliberate selection to change organism
Price 1995:39–40	Natural selection to human-modified environment	Inadvertent (unconscious) artificial selection	Intentional (conscious) artificial selection
Ladizinsky 1998:115–7	Unintentional selection = automatic (Harlan) = unconscious (Heiser)		Intentional selection
Zohary, Tchernov and Horwitz 1998:129	Unconscious, automatic selection to anthropogenic environment		Conscious selection (selective breeding)

Selection Pressures on Plants Cultivated by Humans in the Neolithic

Cyril Darlington revived Darwin's concept of unconscious selection in 1956, but later chose to refer to the selection pressures that followed automatically from systems of tillage, sowing, and harvesting as "operational selection" (Darlington 1963: 156). Those same pressures were equated in a quite misleading way with Darwin's "unconscious selection" in Darlington's last major text *The Evolution of Man and Society* (1969), thereby confusing the issue for many subsequent researchers.

When Jack Harlan, J. de Wet, and Glenn Price (1973) reviewed the domestication of cereals, they adopted the contrasting headings "Automatic Selection" and "Deliberate Human Selection." It became apparent that a large number of the changes occurring in the transformation of wild to domestic forms were of this automatic character: they would not have been obvious to early human farmers, or if noticeable could not be controlled given contemporary agricultural techniques or equipment. In their view, the first time humans sowed seed that they had harvested, automatic selection began for any modifications to the plant and its seed head that enhanced seed recovery and competition in the new environment of cultivation (Harlan, de Wet, and Price 1973: 314). These modifications included unconscious selection for mutant nonshattering seed heads, which in the wild are quickly eliminated for lack of fitness, as well as loss or reduction of seed dormancy, a natural insurance policy for areas with erratic rainfall. Linked to dormancy is the array of seed appendages (glumes, lemmas, and paleas) containing germination inhibitors and mechanically adapted as seed implantation devices in the wild cereal. Automatic selection led to their reduction—it is hard to imagine a Neolithic farmer laboriously isolating seeds from individual plants showing smaller awns. The resowing of previously harvested seed en masse automatically favored plants that produced more seed. In the case of maize, sorghum, or pearl millet, plants that produced a larger inflorescence with larger seed were probably both deliberately selected at harvest, and their genotypes automatically selected through their seedlings' greater vigor when resown. Harlan, de Wet, and Price concluded that in cereals

> Automatic selection pressures are multiple and interlocking, all leading in the same direction and mutually reinforcing, i.e. toward better seed retention, more determinate growth, larger inflorescences, larger seed, loss of dormancy, and so on. (1973: 323)

In contrast, they were of the opinion that

> Deliberate human selection pressures are more absolute and capricious, involve color, flavor, texture, uses, culinary and nutritive values, curios and freaks, and vastly increase the diversity of cultivated races.

The Australian agronomists, Colin Donald and John Hamblin (1983: 100) argued that Darwin's "unconscious" selection was a misnomer, and that the herdsman or cultivator still makes a conscious choice even if he does not have a predetermined standard. They classified this uninformed selection as "nonspecific selection by man." More radically still, they proposed that Darlington's (1969) "unconscious selection by the cultivator" be renamed "natural selection for adaptation to agriculture," thereby removing the category of unconscious–automatic–operational selection from being a form of artificial selection to constituting a special case of natural selection. This new terminology was adopted by Peter Hanelt (1986: 193) but not by Charles Heiser (1988: 80) who credited "unconscious" selection for the early changes to domesticated seed crops.

Examination of the role of unintentional human-influenced selection in the domestication of cereals was extended to several other plant types in the 1980s. The botanist Daniel Zohary fully recognized that some of the divergence encountered in domesticated plants is "the outcome of deliberate, conscious selection by the cultivator" (1984: 579), citing the diversity of form and function in maize and brassicas. However, he went on to say

> More significant still are the automatic, unconscious selection pressures caused by the transfer of plants from their native wild environments into new and contrasting systems of cultivation, the use of different methods of maintenance, and the various ways of utilization of crops. (Zohary 1984: 579)

He stressed the role of the organ of interest to humans in automatically producing different and contrasting selection pressures. Thus, plants grown from seed for their seed (e.g., cereals) must retain their reproductive system intact, whereas plants grown from seed for their fleshy fruits (e.g., tomatoes), can show much greater variation in seed set. More drastic changes in fertility occur in plants that are vegetatively propagated (i.e., cloned) for their fruit. The basic reproductive elements have to be retained for the fruit to develop, but they can become infertile or fail to form seed at all (Zohary 1984: 582–583). Many crops grown by vegetative propagation for their leaves, stems, or tubers cease to flower (e.g., garlic), and may display bizarre chromosomal states. Although the cloning of an exceptional wild fruit tree to perpetuate a desirable type of fruit is an act of conscious selection, this single-step domestication event sets in train a slow process of unconscious selection for reduced fertility, which may be accompanied by automatic selection of mutants (e.g., parthenocarpic figs and pears, hermaphroditic grapes) that can sidestep the problems of fruit set (Zohary and Hopf 2000: 144–145). Clearly, sterility and chromosomal abnormalities were not the intended consequences of the cultivators. Their goal was good fruit in abundant quantities.

Peter Hanelt (1986: 190) similarly stressed the different paths of domestication in plants, depending on the parts desired, and the propagation methods. He was reluctant to accept generic lists of domestication traits, the "Domestikationssyndrom" proposed by Karl Hammer (1984: 30) for crops grown by seed- and vegeculture, because no class of genetic modifications were common to all domesticated plants. Instead, he defined the domestication syndrome as "the crop specific combination of characters which had been evolved as response to natural and artificial selection processes under growing conditions essentially shaped by human activities" (Hanelt 1986: 192). From this, it can be seen that he treats unconscious selection as a form of natural selection. His views on the trends in selection are especially relevant: domestication of seed crops from the wild was initially produced by "natural selection for adaptation to agriculture," and only later in Classical times were artificial mass selection methods applied. In horticultural crops, he believed that mass selection was dominant from the outset, with occasional individual selection of unusual or striking forms (Hanelt 1986: 193).

Gideon Ladizinsky (1998) introduced yet another synonym in his recent text: "unintentional" selection, which with "intentional" selection are forms of "human" selection, as distinct from "natural" selection and "disruptive" selection (Ladizinsky 1998: 113–121). He, too, envisages intentional selection as progressively playing a more important role after the first stages of domestication, especially in seed crops. But just how soon were human cultivators in a position to practice intentional selection, either with nonspecific objectives for what was simply perceived as good stock, or with a deliberate improvement in mind? How easy was it for farmers to separate the grains of wheat or seeds of lentils from the "best" individual plants, when crops were often mixed. How could pure seed be harvested from fields that contained weeds that mimicked the growth habits and timing of the crops?

The archaeobotanical evidence suggests that in the seed crops (cereals and legumes), once morphological domestication occurred, there was little change for several millennia. This may be evidence that unintentional selection had been the dominant mode, because we might anticipate more rapid change with humans in control. In the lentil, domestication occurred by the late 8th millennium B.C.E., but seed size hardly increased beyond that of the wild progenitors for 1,500 years (Ladizinsky 1998: 116; Zohary and Hopf 2000: 100). Maria Hopf described the long-term picture of size increase in an earlier work:

> Prehistoric seeds are normally considerably smaller than the modern high-bred varieties we are familiar with today. ... If one would choose samples from later periods, up to Roman or even early Medieval times, the difference in size would not be much less. One may say, after the various crops once had been established, little further development can be noticed in general until the beginning of modern plant-breeding. (1986: 36)

This was true of several other pulses (Zohary and Hopf 2000: 94). Because the majority of early Near Eastern legumes were self-pollinating, the argument that mutations favoring larger seeds would have been swamped before becoming established cannot apply. It seems that such pulses were not the object of deliberate selection for larger size for many millennia.

We should not expect this prolonged period of unintentional selection to be evident in horticultural crops in which individual seed or clone selection was made by the cultivator. However, the first fruit trees domesticated in the Near East do not appear until the 4th millennium B.C.E., spreading rapidly in the Early Bronze Age (Zohary and Hopf 2000: 142, 174; Zohary and Spiegel-Roy 1975). There is no evidence for domesticated vegetables before that time (Leach 1982; Zohary and Hopf 2000: 250). There is thus a period of some four millennia when unintentional selection appears to have played a more important role in domestication of cereals and pulses in the Near East than any form of deliberate selection. During that long period, very little progress was made toward the domestication of horticultural crops. Significantly, the fruits that required grafting as a technique of cloning, did not appear to be domesticated until the 1st millennium B.C.E.

Selection Pressures on Animals Raised by Humans in the Neolithic

As with plants, there has been active debate since the 1960s over the definition of domestication as it applies to animals, and how the various selection pressures should be categorized. A recent contribution has been Edward Price's division of artificial selection into intentional (conscious) and inadvertent (unconscious)—both result from directed selection. He provides an example of intentional selection for large breast size in turkeys, in which inability to copulate naturally has been inadvertently selected as well (Price 1999: 253). Selection resulting from environmental modification by humans (e.g., conditions of captivity) is treated by Price as a special form of natural selection, comparable to "natural selection for adaptation to agriculture" discussed above. In contrast, for Eitan Tchernov and Liora Kolska Horwitz (1991: 57) the unconscious selection pressures imposed by the anthropogenic milieu on early domesticated animals, are still a type of human selection.

There is greater agreement over the changes that domestication produced, than over typologies of selection. Several key changes in morphology have come to characterize the early domestication of animals, particularly the earliest domesticates of Eurasia: the dog, sheep, goat, pig, and cattle. They form part of the criteria of domestication listed and discussed in all the standard texts, and in the classic works on the domestication of animals (reviewed in Leach 2003). The criterion most used by archaeozoologists to distinguish an early domesticate is a significant decline in body size and robusticity relative to the wild progenitor. In addition, the cranial capacity fell while the facial region of the skull changed its proportions to shorter and broader, often affecting the spacing of the teeth, and eventually their size. Sexual dimorphism declined and in horn-bearing animals there was greater diversity in shape and size of these structures.

Less is known about the selection pressures that produced these effects than is the case with the domestication syndromes in plants. For example, the size change in animals has been attributed to combinations of genetic, epigenetic, and developmental factors. Restricted gene flow following isolation of herds, and unconscious selection of genotypes capable of surviving nutritional stress, crowding, and disease are examples of genetic changes, whereas malnutrition affecting individual phenotypes provides an example of a developmental influence. Relaxation of natural selection may be invoked, as in the argument that small individuals survived to pass on their genotype when their natural predators were removed. Deliberate selection for small docile animals and culling of large aggressive males has also been proposed. This may have produced a string of unforeseen consequences, often described as heterochronic, because they disrupt the developmental processes, for example allowing the animal to reach sexual maturity while retaining many juvenile characters. The range of explanations offered since the 1950s is large; however, over the past decade, less emphasis has been placed on deliberate selection for breeding stock than on the unintentional consequences of changes in environment, diet, mobility, and of the removal of natural selection pressures. The mechanisms by which heterochronic changes occur are the subject of the most recent research. For example, differential stress tolerances affecting the thyroid hormone are proposed as a simple biological mechanism for the domestication traits in early dogs (Crockford 2000: 11). Thus, for the first wave of domesticates, many researchers now believe that the powerful selection pressures came from adaptation to new environments, in both animals and plants rather than from human breeding decisions.

According to Sándor Bökönyi, the reduction in size of cattle, sheep, and goats continued for several millennia in southeastern Europe. A substantial drop in cattle wither height marked initial domestication, but the downward trend persisted, reaching a nadir in the Iron Age cultures just prior to the Roman imperial period (Bökönyi 1974: 115). From his extensive experience as an archaeozoologist, Bökönyi proposed that animal husbandry within the context of domestication consisted of two phases. The first he described as *animal keeping*, "a rather instinctive activity lacking both conscious selection and proper feeding techniques." The animals, thus, kept were very variable in their characteristics, were of primitive types, and tended to be much smaller than their wild progenitors. The second phase was *animal breeding*, "based on conscious selection of suitable animals as parents and appropriate feeding practices." In this more advanced phase, species were represented by several breeds, and they were generally larger and more productive, although miniature breeds might also be maintained (Bökönyi 1984: 10). Selection did occur in the Neolithic, in the sense that some animals were selectively prevented from breeding by castration or early culling, but this was not conscious breeding intended to increase productivity. In Bökönyi's view, conscious breeding began in Mesopotamia in the 3rd millennium B.C.E. in some special sheep flocks, but was not introduced to Europe until the 1st millennium B.C.E., by the Scythians and the Greeks. In Italy, morphological differences in dogs are consistent with "accidental cross-breeding" from the Neolithic until the Iron Age, when there are signs of selection for large size (Mazzorin and Tagliacozzo 2000: 150–155). Lap dogs first appeared in the Roman era and were clearly dependent on conscious breeding decisions for survival, because they risk death during whelping when allowed to breed promiscuously.

Selection in Classical Animal Husbandry

Deliberate or conscious breeding for improvement was the essence of Darwin's methodical selection, and there are many commentators on the classical texts who have interpreted them as evidence of such goal-directed practice. Using both texts and archaeozoological evidence, Bökönyi (1984: 10) argued that the Romans further developed the "knowledge of conscious animal breeding" and applied it to all the domestic animal species, resulting in the formation of "improved breeds." These were transported to the Roman provinces. How advanced then, was Greek animal breeding? A recent work on Greek agriculture stated that Aristotle, the author of *Historia Animalium,* was conscious that "like cultivated plants, domesticated animals have developed characteristic species by means of more or less deliberate breeding" (Isager and Skydsgaard 1992: 85). Aristotle's plentiful advice on how to select animals for breeding was cited in support of an assumption that "a very deliberate effort with regard to breeding has taken place, including the purchase and transport of breeders aiming at an improvement of the stock" (Isager and Skydsgaard 1992: 96), an effort intensified in Hellenistic and Roman times when very high prices were paid for breeding stock.

It is important not to equate classical animal breeding with that practiced by Darwin's contemporaries, without first examining the types of selection involved. As Denison Hull (1964: 22) said in reference to Greek hounds, "It is tempting to give up the very word 'breed,' on the ground that the Greeks kept no studbooks and had no systematic way of recording pedigrees," but he gave three reasons for accepting the existence of breeds. The first was the recommendation attributed to the late Greek poet Oppian, not to cross breeds, but to keep them pure; the second was Xenophon's warning that the milk of (what Hull interpreted as mongrel) foster mothers was not good for purebred puppies; and the third was the Greeks' high opinion of good breeding in humans as evidenced by the recitation of long male line genealogies. We know from Aristotle that a horse breeding establishment was "not considered perfect unless horses mount their own progeny" (Peck 1970: 321). But does the Greek obsession with eugenics, in its original meaning of "good breeding," mean anything more than the attempt to maintain the good qualities of the sire in a few elite strains?

The dozens of dog "breeds" known to the Greeks, which Hull (1964) went on to enumerate, were remarkable for the fact that their Greek names were derived from regions and peoples stretching from the

Iberian Peninsula to Egypt and Central Asia. This must constitute strong evidence that most of these so-called breeds were actually regional types, equivalent to landraces in plants, adapted to the climates of those regions and to their roles in human society, whether for stock guarding, warfare, hunting, or as food. In the case of the dog type that Xenophon referred to as Vulpine, his practical knowledge of selective breeding might be judged from his statement that this originated in a cross between a dog and a fox, and that the nature of the parents had become fused in their descendants (Marchant 1968: 377).

Although Bökönyi was a strong believer in the advanced skills of the Roman animal breeder, he could not contemplate a Roman canine world consisting of numerous pure breeds:

> Even in such a highly civilized society such as the Roman, the majority of dog breeds were not kept separately. With the exception of the luxury dogs (which even today are kept isolated from other breeds), all the watchdogs, herding dogs and pariah dogs had excellent opportunity to interbreed with each other at will. (1984: 92)

In an empire as large and as interconnected as the Roman one, there were many opportunities and much commercial motive to transport regional types of dogs, horses, cattle, sheep, pigs, pigeons, and other species to new locations where they might prove more useful or profitable than existing stock. This does not automatically imply that classical farmers were practicing stock improvement through the sort of program of out-crossing followed by inbreeding that, two millennia later, created the 19th-century breeds so laboriously tracked in the stud books. In fact, one of the most authoritative sources on Roman farming, Kenneth D. White (1970: 500), considered that in the case of cattle, there was almost no trace of selective breeding as is now understood. Certainly, the Romans made long lists of desirable traits to be looked for when purchasing animals for particular uses. They were also aware of the influence of local soils and climate. Columella (Forster and Heffner 1968: 235, 237) even reported on an experimental crossing of domestic sheep in Spain with wild North African rams. Varro noted perceptively that ram lambs and buck kids to be raised for breeding stock should preferably be selected from the offspring of mothers that bear twins (Hooper 1967: 343, 347). However, the Roman writers could not agree on the contributions of the sire or dam to the offspring (White 1970: 285, 501), and in the case of rams were convinced, following Aristotle, that the color of their tongues could predict the fleece colors of their lambs

(White 1970: 308). They believed that the sex of future calves could be predicted from the side of the cow on which the bull dismounts (Hooper 1967: 375). In White's view, "there is a world of difference between the careful selection of the animal most suited to a specific purpose, whether economic or sacrificial, and selective breeding, which entails the selection of sires and dams according to the stud-book" (1970: 312). Although the Romans valued choice, their deficient knowledge of heredity confined them to the maintenance of what had originated as regional types, rather than combining the best qualities of several types into new, improved breeds. At best, they practiced Darwinian unconscious selection.

Selection in Classical Agriculture and Horticulture

White believed that greater progress had been made in the classical world with plant than with animal breeding. The writings by Plato's pupil Theophrastus, in the 4th century B.C.E., show considerable understanding of the problems of maintaining good types of fruits (Einarson and Link 1976: 67–69). In his experience, cultivated fruit trees raised from seeds were generally inferior, and might even revert to wild forms. He perceived the cause as weakness of the seeds, resulting from too much feeding under cultivation, whereas we now know that many fruit trees are out-breeders and genetically highly heterozygous (Ladizinsky 1998: 43; Zohary and Hopf 2000: 144). Their offspring seldom resemble either parent; hence the recognition by the Greeks that asexual propagation from cuttings, offshoots or the more difficult grafting was necessary to reproduce fruit quality. But Theophrastus also believed that cultivated varieties can change to better forms under the influence of location, citing the sweet pomegranate that developed in Egypt and the "stoneless" [seedless?] pomegranate of Cilicia (Einarson and Link 1976: 69). We would attribute such a change to a mutation rather than to the influence of the environment. For the Romans, the desire to try exotic varieties of their traditional fruits posed considerable risks to the grower. For the grape vine, Columella warned that "experience teaches that to every region its own variety is more or less suited" (Ash 1960: 229). Indeed, he knew of some varieties that had "so far departed from their peculiar character, through a change of place, as to be unrecognizable" (Ash 1960: 251).

How did selection occur in classical seed crops? Theophrastus's advice was to select seed for resowing from plants of prime quality (Hort 1961: 79), the same principle applied to animals. Mixing local and foreign varieties of a seed crop was risky, because they might need to be sown at different times, and germinate unevenly (Hort 1961: 191). The Romans had good reason to worry about the mixing of varieties, and the unexpected results obtained when new varieties were tried. For a start, they had both winter and summer types of wheat, and recognized within them several types more or less suited to different rainfall regimes and soil moisture levels. Writing some two generations after Varro, Columella noted the deterioration of all types of wheat to a light inferior variety when sown for three years in wet ground (Ash 1960: 151–153). He agreed with the poet Vergil that degeneration was almost inevitable, even after handpicking the best heads from the threshing floor for future sowing. This apparent degeneration of varieties was almost certainly the result of seed mixture—after a few years, the variety within the mixture that was most suited to the conditions became dominant (White 1970: 188). The Romans did not have the facilities to ensure pure seed. Another method of selection was to choose the grain that settled to the bottom of the sieve, on the grounds that it would be the largest and heaviest. Although this can be seen as conscious selection for size and density, the Romans would not have been aware that under this method, "there is usually a compensating decrease in the number of grains per plant" (Donald and Hamblin 1983: 110). As with animal husbandry, the Roman farmer was limited to selection for maintenance of what he perceived as good strains; however his efforts were constantly offset by natural and unintentional selection pressures that he accepted fatalistically.

Selection in the Third Stage of Domestication

Archaeozoological evidence from Central Europe, after the collapse of the Roman provinces, portrays a return to the small primitive breeds that prevailed in the Iron Age. Bökönyi (1974: 86) believed that selective breeding effectively ceased until the late Middle Ages. Most commentators now place the onset of artificial selection, in Darwin's sense of methodical breeding for improved characteristics (whether ornamental or functional), somewhat later. For example in England, Juliet Clutton-Brock (1998: 186) placed the start of this stage of deliberate selection in the mid–18th century. It can be traced to the 17th century, when Arab sires were imported to improve the performance of racehorses. The sires were mated with their own progeny and their progeny in turn, for four generations or more in a sequence of back-crosses intended to increase the amount of "noble" blood to 95 percent (Wood and Orel

2001: 48). Although some rejected the process of breeding "in-and-in" as incest, and thus against divine rule, a justification was found that it was simply returning a degenerated type back to its original perfection (Wood and Orel 2001: 39).

The principles were then applied to sheep, cattle, and pigs. For pig breeders, the desirable new traits were to be found in the paedomorphic Asian pigs. Whereas the Romans would have attempted to import the new variety and maintain its purity, in this "modern" era, controlled crossbreeding was seen as the key to improving all the old regional forms. An explosion of pig varieties resulted, of largely unstable character, until breeds were standardized by the stud keepers in the later 19th century (Malcolmson and Mastoris 1998: 70–73). One vital shift in landholding was a prerequisite for the maintenance of true breeds. Whereas from Neolithic times, most pigs had spent long periods feeding in forests and on waste ground where their breeding could not be controlled, land enclosure now meant that they had to be confined to pens and sties. Here, both their diet and breeding could be effectively managed. Interbreeding with feral pigs was no longer a possibility. Similarly, in sheep, once they no longer grazed as unfenced flocks on commons or the open hillsides, selective breeding for improvement could be more effective. In this case, the new forage crops and new rotation sequences developed in the 18th century were essential (Wood and Orel 2001: 36).

Such planned programs of crossbreeding and inbreeding, coupled with progeny testing, characterized the third stage of domestication until recently. As knowledge of genetics developed, the risks of inbreeding became better understood. Nevertheless, it has become obvious that as the problems of genetically transferred conditions have been brought under control or at least understood, the dangers of breeding for highly specialized function have become increasingly evident, for example in the form of behavioral abnormalities, structural disorders, and difficulties in unassisted mating and birthing. These occur in addition to the undesirable traits that are genetically linked to favored characters. Gene mapping and studies of quantitative trait loci are now revealing the nature of these linkages, and genetic engineering is often portrayed as a means of overcoming the problems they create.

Selection in the Fourth Stage of Domestication

As the science of genetics matured in the 20th century, both the knowledge base and the necessary laboratory technology advanced to the point at which selected pieces of DNA from one organism of one species could be inserted into the chromosomes of another (Royal Society 1998: 22; Tudge 2000: 233–235). The boundaries that had hitherto restricted the exchange of DNA (by sexual means) to the level of species, or exceptionally genus, were circumvented. Phenotypic selection for desirable qualities gave way to genotypic selection, not just for the genotype as a whole but for individual genes. Selection at this fourth stage is predicated on genomics, the knowledge of the function of the genes, and their positions on the chromosomes. Thus, the locus of selection has shifted from the farm or garden to the laboratory, and the agent from the plant or animal breeder to the molecular biologist (Rader this volume).

Criteria of Animal and Plant Domestication

In the 1980s, the concept of a "domestication syndrome" applicable to all domesticated plants was rejected (Hanelt 1986: 192). Instead, combinations of nonwild characters repeatedly seen in members of the same types of crops, such as tillering seed crops, or pome-bearing fruits, are cited as markers of their particular paths to domestication. The criteria of domestication in animals should be similarly restricted. They are frequently discussed as though generally applicable, when as yet their existence is only well documented in a selection of Old World domesticated mammals. It is obvious that the criteria are a mixture of traits, some of which we know, from the archaeozoological evidence, appeared in the first few millennia of the Holocene era, whereas others including pellage characters and changes in sexuality necessarily have a more uncertain chronology. The set of criteria is the end product of many millennia of domestication, during which the dominant selection pressures underwent significant changes. We should not assume that the criteria appeared simultaneously. Some may be typical of the first stage of domestication, others much more recent and resulting from goal-directed selective breeding.

Stages of Biological Domestication and Associated Types of Selection

I have divided domestication into four stages, in each of which additional selection pressures were brought into play (see Table 2). In the first stage two types of unintentional selection prevailed, as the affected species adapted to human-modified environments and early

Table 2. Stages of Biological Domestication and Associated Types of Selection

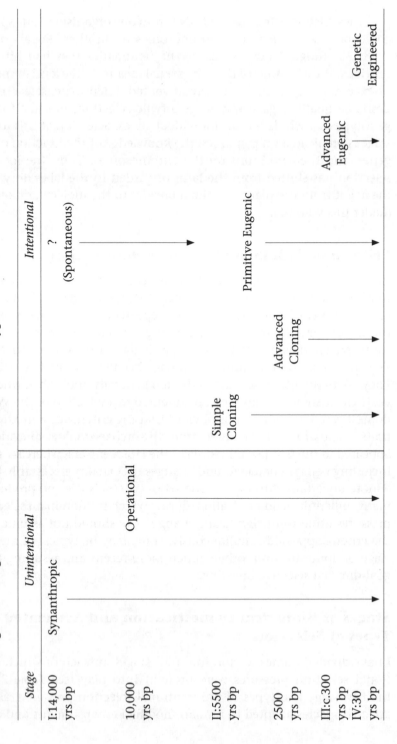

Stage	Unintentional	Intentional
I:14,000 yrs bp	Synanthropic	? (Spontaneous)
10,000 yrs bp	Operational	
II:5500 yrs bp		Simple Cloning
2500 yrs bp		Advanced Cloning
III:c.300 yrs bp		Primitive Eugenic
IV:30 yrs bp		Advanced Eugenic
		Genetic Engineered

agricultural practices. I have named these *synanthropic* and *operational* selection.

Synanthropic selection occurs where a species is attracted to the highly modified environment that surrounds humans. It may be a constructed environment such as a collection of mud-brick houses or it may be a natural rock shelter, but it must provide shelter, protection from predators, and access to foodstuffs. Initial behavioral modification, such as diminished flight response and begging, eventually leads to genetic modification through differential selection, but only if the humans occupy the site for long enough for breeding to occur in the affected species, or if (like the dog) it is capable of traveling to new sites with humans. Humans are a special case of synanthropic domestication, adapting biologically to their own environmental modifications (Leach 2003). The Greek word *synanthropeuonema* was used by Aristotle to categorize creatures that live with humans. He noted that pigs, dogs, domestic birds, and insects sharing human housing can mate at any time of the year "owing to the shelter and the plentiful food they get by their association with man" (Peck 1970: 117). Certain weedy plants thriving in yards and domestic middens might be considered synanthropic self-domesticates, comparable to the house mouse and house sparrow. In zoological terms (O'Connor 1997: 151), synanthropic species may exhibit mutualism (in which the relationship is beneficial to both humans and the cohabiting species, as with the dog), commensalism (beneficial to one and neutral to the other, as with the sparrow), or contramensalism (beneficial to one at the expense of the other, as with the mouse). The first synanthropic mammal domesticate was the dog, which evolved from a species of wolf before full human sedentism. The wolf's habit of traveling in a pack can be seen as a preadaptation for migration with a human group. The dog became the only synanthropic domesticate of many hunter-gatherer and forager societies.

Operational selection occurs when a species undergoes genetic change in the course of adapting to human farming or gardening operations, and when these changes are not the intended consequence of deliberate selection by humans. In plants, not all forms of cultivation lead to operational domestication; so there is a period of unknown length at the Neolithic interface when predomestication cultivation cannot be ruled out. Not only are farmers and gardeners unaware of many of the changes that become genetically expressed in crop plants as a result of unintended operational selection, but the process can also produce domesticated "weeds." Undesirable species sharing the field or garden plot can be equally affected by the operations of tillage, sowing, weeding, and harvesting. They may adapt in similar ways to the crop, mimicking their morphological changes, and ensuring continued representation in the seed set aside for resowing. Some of the worst "weeds" of field and garden crops are operational domesticates.

Although the concept of operational selection was developed and used by botanists (Darlington 1963: 156; Heiser 1988: 77–78), it is applicable to animals and possibly other organisms. If reduction in body size in early goats and sheep is not the result of deliberate selection of small animals for breeding, but of unintentional selection of animals adapted to poorer nutrition, or the removal of natural selection pressures (such as predation) which eliminates smaller individuals, then the operations of penning and herding might be said to have led to operational domestication of these early farmed animals. Subsequently, sheep and goats were subject to intentional selection. A case might be made today that drug-resistant microorganisms are a special category of operational domesticate, comparable to herbicide-resistant "weeds."

In the second stage of domestication, intentional although unconscious (in the Darwinian sense) selection pressures were added as humans tried to maintain desirable types of domestic animal outside their original area of adaptation—I have called this *primitive eugenic* selection. In plants *primitive* and later *advanced cloning* techniques selected desirable varieties of fruit and vegetable.

Primitive eugenic selection takes its name from the Greek term *eugenes* meaning "of good stock" or "of high breeding" (Peck 1965: 19), applied to both humans and economically important animals, such as sheep, donkeys, and horses. Because the Greeks considered that both desirable and undesirable qualities were transmitted through the male line, it was the male of the species that was subject to the most active selection (Isager and Skydsgaard 1992: 101). In both animals and humans, pedigree was founded on male inheritance. Frequently, the goal of selection was to maintain the qualities of local varieties transported beyond their place of origin. The Romans extended primitive eugenic selection to crops, endeavoring to isolate the strain and to handpick the best seed heads.

The antiquity of early eugenic selection is uncertain. Arguably, it arose from spontaneous selection of a type driven by the exercise of personal preference in a spur-of-the-moment act. The choice of a pup from a litter or of a ram lamb for a sacrifice might or might not have affected the gene pool of particular domestic populations. The long-term effects of such choices did not motivate the selection, however, and therefore spontaneous selection, although intentional, cannot qualify as a breeding strategy. But repeated cultural consistency in the criteria of choice, such as for animals with particular coat colors, provided a direction to the process that accelerated phenotypic differentiation, the development of local varieties and ultimately genetic isolation and drift. Recognition of groups of animals that shared particular traits led to their naming as a type. The desire to perpetuate that type gave rise to the first strategies of selection that qualify as eugenic.

Selection through the application of simple cloning techniques was exclusive to plants. In this process, propagation through sexual means (fertilization followed by seed development) was replaced by vegetative propagation. Many plants can be reproduced through the detachment and planting of rhizomes, bulbils, corms, or suckers; or through the rooting of cuttings taken from stems or shoots. In the absence of sexual reproduction, these plants are of course identical to the parent plant. The advantage of a plant clone is that desirable qualities can be perpetuated, in theory ad infinitum. However somatic mutations and damage to cells from viral infections can lead to genetic change and often a slow deterioration of health in a cloned variety. Several early tree and vegetable crops were subject to this type of vegetative propagation, such as olives, grapes, and garlic.

Selection through advanced cloning took longer to appear. It was applied to plant varieties that did not readily root from cuttings but required the grafting of the desirable variety on to a suitable rootstock. A high level of expertise is needed to match the budwood to the stock (which may be of a different but related species or genus). Grafting was common in Roman times and was used on orchard trees such as apples. In the 20th century, another advanced cloning technique developed: tissue culture, which is performed under laboratory conditions.

In the third stage of domestication, methodical selection pressures were added as breeders learned how to improve and "fix" breeds—this can be called *advanced eugenic* selection. It was built initially on the growing knowledge of heredity in the 18th century, in particular the benefits of judicious hybridization or crossing followed by inbreeding to fix desirable characteristics. It required careful record keeping, progeny testing, and rigorous culling. Darwin and Mendel drew on the results of breeding experience to formulate the mechanisms of evolution and rules of heredity, but it was not until the discovery of genes that the reasons for particular negative outcomes of eugenic breeding became apparent. In animals, this form of selection still depended on sexual reproduction. However, in their search for variation, plant breeders learned how to accelerate the rate of mutation and induce polyploidy that could render formerly sterile hybrids fertile, creating new species in the process.

In the fourth stage of domestication, *genetic engineering* advances the goals of varietal improvement through direct selection of desirable genes and their insertion into the chromosomes of what may be unrelated species. This form of selection draws on the laboratory techniques pioneered by plant geneticists, such as the transfer of genetic material by vectors, and tissue cloning. It differs from earlier types of selection in its concentration on particular genes instead of whole phenotypes or genotypes, and it is no longer constrained by the traditional intrageneric or intraspecific boundaries of sexual reproduction.

Under a four stage subdivision of domestication, the differing degrees of modification and human control can be easily accommodated. A pedigree toy spaniel represents the product of stepwise selection pressures on dogs that started in the first stage with facial shortening perhaps 14,000 years ago, shortening of the limbs in lap dogs during the second stage 2,000 years ago, and in the third stage selection for a flat face over the last 100 years. The initial facial shortening was not under the control of humans, whereas the maintenance of the mutations that gave rise to Roman lap dogs was controlled to the extent of preventing mating with larger dogs. The recent creation of the English toy spaniel breed was by genuine goal-oriented artificial selection. Future cloning of particular pet dogs is already offered commercially in the fourth stage (see Rader this volume).

Humans, Domesticated and Wild?

The case has already been put that humans became biologically domesticated as sewn clothing and substantial housing, changes to a softer diet, and reduction of mobility relaxed the selection pressures that had operated for most of the Paleolithic period. The robust phenotype gave way to more gracile skeletal morphologies, and humans displayed several of the traits that occurred in their domesticated animals (Leach 2003). Both humans and animals were modified by selection pressures that were not under conscious control by humans. Thus, many domestic species experienced changes during the first stage of domestication that operated and were experienced in ways not dissimilar to changes produced by natural selection.

In the second stage of domestication, humans were more aware of their power to achieve change in their domestic animals through selection. To a very limited extent, they applied the same principles to themselves. Through concern with genealogy and lineage, leaders of many human societies attempted to maintain their "noble" bloodlines by careful selection of marriage partners. By the time the third stage began, human control over animals seemed paramount, although every step to increased productivity was accompanied by unforeseen consequences that had to be culled (e.g., Stanley 1995: 67). In the 21st century, anxiety over the power of the genetic engineer to alter human genotypes to a similar extent as in animals is becoming widespread. Nevertheless, prevention and correction of human disorders is a significant and generally approved goal in the fourth stage of domestication, in contrast to the notorious eugenic experiments on humans in the third.

In the 21st century, many screening tests are available, including sperm sorting, which facilitates sex selection during in vitro fertilization, preimplantation genetic diagnosis (PGD) to ensure that only healthy embryos are implanted, and prenatal screening for genetic disorders such as Down syndrome. The latter may lead to pregnancy termination, although not invariably. Prenatal sex determination and subsequent abortion of female fetuses has been implicated in increasing gender imbalances in both Chinese and Indian populations. Demographically, this practice is having a greater impact that the screening of oocyte donors in the United States according to measures of intelligence, appearance, and "race." It is not difficult to find analogues for these selection procedures in third- and fourth-stage animal breeding. However, the costs of most modern reproductive technologies restrict them to a very small proportion of the First World and therefore minimize their evolutionary impact on the human species as a whole. Until the majority of humans are screened as parents, embryos, or fetuses, humans will remain less artificially selected (in a eugenic sense) than their domestic animals.

This does not mean that the unconscious selection pressures that contributed to human gracilization in the first stage of domestication are any less significant now than 10,000 years ago. In fact, the 20th century saw the disappearance of the last humans with relatively robust phenotypes, as technologically modified environments spread to the remotest regions of the earth.

As domestication proceeded, those human groups that did not make the transition to agriculture were increasingly seen as alien. The Greek attitude to domestication is particularly revealing. Aristotle saw no problem in discussing humans as a class of animals. When discussing wildness and tameness, he remarked that some animals (like the leopard) are always wild, some can be quickly tamed (like the elephant), and that "any kind of animal which is tame exists also in a wild state, e.g., horses, oxen, swine, men, sheep, goats, dogs" (Peck 1965: 17; cf. Peck 1961: 89). The word used by Aristotle for "tame" was *hemeros,* a word that Theophrastus applied to cultivated (i.e., domesticated) plants. Theophrastus made the significant comment that "Man, if he is not the only thing to which this name is strictly appropriate, is at least that to which it most applies" (Hort 1968: 29). If civilized humans epitomized tameness, who were the men that Aristotle said existed in a wild state? They were not the nomadic barbarians, against whom the Greeks judged their own degree of advancement, but savages who neither tilled the soil nor possessed houses. Some, according to Herodotus, were even man eaters (Godley 1963: 307, 393–395). The word applied to them, *agrios,* was precisely that used for undomesticated animals and wild plants (Powell 1938: 3).

Charles Darwin's opinions on the robust Yanama Fuegians that he met in 1832 were couched in similar language: "I could not have believed how wide was the difference, between savage and civilized man. It is greater than between a wild and domesticated animal" (Barrett and Freeman 1986: 179). He went on to conclude that they "cannot know the feeling of having a home, and still less that of domestic affection" and that their "skill in some respects may be compared to the instinct of animals" (Barrett and Freeman 1986: 185). Many other examples might be cited of derogatory comments made by people whose physique was no longer adapted to climatic extremes, about human groups whose continued robusticity was essential for survival. But they failed to recognize that their own domesticated state was merely a response to a modified environment. The robust Fuegians had not been "softened" by the environment of domestication that accompanied sedentism and agriculture, but that did not mean that they were therefore "wild" humans, trapped in or degenerated into a state of savagery. Both the Selk'num and the Yamana Fuegians possessed domesticated dogs, and the Yamana dictionary compiled by the Rev. Thomas Bridges contained some 32,000 words, many attesting to the complexity and frequency of their social interchanges, including 61 for categories of kinship (Bridges 1948: 529–532; Bridges 1987/1933: xvii). And contrary to Darwin's belief that they were homeless, the word used for a dwelling, however insubstantial or temporary, was precisely the term that described a group of kin, whether household or tribe (Bridges 1987/1933: 48). Just as it was for Darwin, the Yamana "home" was where the family gathered.

For human groups, therefore, biological domestication was an unforeseen consequence of selection pressures associated with the progressive change of lifestyles from the end of the Pleistocene era. As the gracile inherited the world, they came to regard themselves as more advanced than the humans whose robusticity was a biological adaptation to harsh environments. There were some dissenters to this view. Since the Greek Cynics, cultural primitivists have warned of the softening effects of civilization (Lovejoy and Boas 1997/1935), without realizing what a disadvantage they would be at if they discarded its technological buffers. Perhaps the question that contemporary primitivists should now address is how quickly they might regain a robust phenotype for the next glaciation.

From this review, the inescapable conclusion is that for 97 percent of the time since domestication processes began, humans have not understood the mechanisms sufficiently to foresee the consequences for the plants and animals that became their focus, let alone appreciate how they themselves might be changed. Today "we are changing Earth more rapidly than we are understanding it," and current changes in global ecology are "fundamentally different from those at any other time in history" (Vitousek et al. 1997: 498). Justifying even more intensive genetic manipulation of domesticated (and yet to be domesticated) species on the grounds of ten millennia of prior experience is both unsupportable and evolutionarily unsound. What control we currently exercise arises from the infant science of genomics (first named in 1987), not the long history of domestication.

Acknowledgements

I am very grateful to Richard Fox and the Wenner-Gren Foundation, and to the organizers, Molly Mullin and Rebecca Cassidy, for the invitation to participate in this symposium. Thanks are owed to Laurie Obbink and Amy Perlow for making it happen, and to all the contributors for their generous input and advice.

References

Ash, Harrison B. 1960. *Lucius Junius Moderatus Columella on agriculture,* vol. 1, *Loeb classical library.* London: Heinemann.

Bajema, Carl J., ed. 1982. *Artificial selection and the development of evolutionary theory.* Stroudsburg, PA: Hutchinson Ross.

Barrett, Paul H., and Richard B. Freeman, eds. 1986. *The works of Charles Darwin,* vol. 2, *Journal of researches. Part One.* London: William Pickering.

Belzile, François J. 2002. Transgenic, transplastomic and other genetically modified plants: A Canadian perspective. *Biochimie* 84: 1111–1118.

Bökönyi, Sándor. 1974. *History of domestic mammals in central and eastern Europe.* Budapest: Akadémiai Kiadó.

——. 1984. *Animal husbandry and hunting at Tác-Gorsium: The vertebrate fauna of a Roman town in Pannonia.* Studia Archaeologica, 8. Budapest: Akadémiai Kiadó.

——. 1989. Definitions of animal domestication. In *The walking larder: patterns of domestication, pastoralism and predation,* edited by Juliet Clutton-Brock, 22–27. London: Unwin Hyman.

Bridges, E. Lucas. 1948. *Uttermost part of the Earth.* London: Hodder and Stoughton.

Bridges, Thomas. 1987 [1933]. *Yamana–English: A dictionary of the speech of Tierra del Fuego,* edited by Ferdinand Hestermann and Martin Gusinde. Buenos Aires: Zagier y Urruty Publicaciones.

Clutton-Brock, Juliet. 1998. The role of artificial selection in evolutionary thought. In *Man and the animal world: Studies in archaeozoology, archaeology, anthropology and palaeolinguistics in memoriam Sándor Bökönyi,* edited by Peter Anreiter, László Bartosiewicz, Erzsébet Jerem, and Wolfgang Meid, 185–189. Budapest: Archaeolingua Foundation.

Crockford, Susan. 2000. Dog evolution: A role for thyroid hormone physiology in domestication changes. In *Dogs through time: An archaeological perspective,* edited by Susan J. Crockford, 11–20. Oxford: BAR International Series 889.

Darlington, Cyril D. 1956. *Chromosome botany.* London: Allen and Unwin.

——. 1963. *Chromosome botany and the origins of cultivated plants.* 2nd edition. London: Allen and Unwin.

——. 1969. *The evolution of man and society.* London: Allen and Unwin.

Darwin, Charles. 1868. *The variation of animals and plants under domestication.* 2 vols. London: John Murray.

Donald, Colin M., and John Hamblin. 1983. The convergent evolution of annual seed crops in agriculture. In *Advances in agronomy,* vol. 36, edited by N. C. Brady, 97–143. Orlando: Academic Press.

Einarson, Benedict, and George K. K. Link. 1976. *Theophrastus. De causis plantarum,* vol. 1, *Loeb classical library.* London: William Heinemann.

Forster, Edward S., and Edward H. Heffner. 1968. *Lucius Junius Moderatus Columella on agriculture,* vol. 2, *Loeb classical library.* London: William Heinemann.

Godley, Alfred D. 1963. *Herodotus,* vol. 2, *Loeb classical library.* London: William Heinemann.

Hammer, Karl. 1984. Das domestikationssyndrom. *Kulturpflanze* 32: 11–34.

Hanelt, Peter. 1986. Pathways of domestication with regard to crop types (grain legumes, vegetables). In *The origin and domestication of cultivated plants,* edited by Claudio Barigozzi, 179–199. Amsterdam: Elsevier.

Harlan, Jack, J. M. J. de Wet, and E. Glenn Price. 1973. Comparative evolution of cereals. *Evolution* 27: 311–325.

Heiser, Charles B. 1988. Aspects of unconscious selection and the evolution of domesticated plants. *Euphytica* 37 (1): 77–81.

Hooper, William D. 1967. *Marcus Porcius Cato on agriculture. Marcus Terentius Varro on agriculture, Loeb classical library.* London: Heinemann

Hopf, Maria. 1986. Archaeological evidence of the spread and use of some members of the Leguminosae family. In *The origin and domestication of cultivated plants,* edited by Claudio Barigozzi, 35–60. Amsterdam: Elsevier.

Hort, Arthur. 1968. *Theophrastus. Enquiry into plants,* vol. 1, *Loeb classical library.* London: Heinemann.

———. 1961. *Theophrastus. Enquiry into plants,* vol. 2, *Loeb classical library.* London: Heinemann.

Hull, Denison B. 1964. *Hounds and hunting in ancient Greece.* Chicago: University of Chicago Press.

Isager, Signe, and Jens E. Skydsgaard. 1992. *Ancient Greek agriculture: An introduction.* London: Routledge.

Ladizinsky, Gideon. 1998. *Plant evolution under domestication.* Dordrecht, the Netherlands: Kluwer Academic.

Leach, Helen M. 1982. On the origins of kitchen gardening in the ancient Near East. *Garden History* 10 (1): 1–16.

———. 2003. Human domestication reconsidered. *Current Anthropology* 44 (3): 349–368.

Leakey, Richard E., and Roger Lewin. 1996. *The sixth extinction: Biodiversity and its survival.* London: Weidenfeld and Nicolson.

Lovejoy, Arthur O., and George Boas. 1997 [1935]. *Primitivism and related ideas in antiquity.* Baltimore: Johns Hopkins University Press.

Malcolmson, Robert, and Stephanos Mastoris. 1998. *The English pig: A history.* London: Hambledon Press.

Marchant, Edgar C. 1968. *Xenophon. VII. Scripta minora, Loeb classical library.* London: Heinemann.

Mazzorin, Jacopo de Grossi, and Antonio Tagliacozzo. 2000. Morphological and osteological changes in the dog from the Neolithic to the Roman period in Italy. In *Dogs through time: An archaeological perspective,* edited by Susan J. Crockford, 141–161. Oxford: BAR International Series 889.

O'Connor, Terry P. 1997. Working at relationships: Another look at animal domestication. *Antiquity* 71 (271): 149–156.

Peck, Arthur L. 1961. *Aristotle. Parts of animals, Loeb classical library.* London: Heinemann.

——. 1965. *Aristotle. Historia animalium,* vol. 1, *Loeb classical library.* London: Heinemann.

——. 1970. *Aristotle. Historia animalium,* vol. 2, *Loeb classical library.* London: Heinemann.

Powell, J. Enoch, 1938. *A lexicon to Herodotus.* Cambridge: Cambridge University Press.

Prakash, Channapatna S. 2001. The genetically modified crop debate in the context of agricultural evolution. *Plant Physiology* 126: 8–15.

Price, Edward O. 1995. Behavioral genetics and the process of animal domestication. In *Genetics and the behavior of domestic animals,* edited by Temple Grandin, 31–65. San Diego: Academic Press.

——. 1999. Behavioral development in animals undergoing domestication. *Applied Animal Behavior Science* 65 (3): 245–271.

Royal Society. 1998. Genetically modified plants for food use. Ref2/98. Available at: http://www.royalsoc.ac.uk/document.asp?tip=1&id=1929, September 15, 2006.

Sax, Boria. 1997. What is a "Jewish dog"? Konrad Lorenz and the cult of wildness. *Society and Animals* 5 (1): 3–21.

Stanley, Pat. 1995. *Robert Bakewell and the Longhorn breed of cattle.* Ipswich: Farming Press.

Strauss, Steven H. 2003. Genetic technologies—Genomics, genetic engineering, and domestication of crops. *Science* 300 (5616): 61–62.

Tani, Yutaka. 1996. Domestic animal as serf: Ideologies of nature in the Mediterranean and the Middle East. In *Redefining nature: Ecology, culture and domestication,* edited by Roy Ellen and Katsuyoshi Fukui, 387–416. Oxford: Berg.

Tchernov, Eitan, and Liora Kolska Horwitz. 1991. Body size diminution under domestication: Unconscious selection in primeval domesticates. *Journal of Anthropological Archaeology* 10: 54–75.

Tort, Patrick. 1996. *Dictionnaire du Darwinisme et de l'évolution.* Paris: Presses Universitaires de France.

Tudge, Colin. 2000. *In Mendel's Footnotes. An introduction to the science and technologies of genes and genetics from the nineteenth century to the twenty-second.* London: Jonathan Cape.

Vitousek, Peter M., Harold A. Mooney, Jane Lubchenco, and Jerry M. Melillo. 1997. Human domination of Earth's ecosystems. *Science* 277 (5325): 494–499.

White, Kenneth D. 1970. *Roman farming.* London: Thames and Hudson.

Wood, Roger J., and Vítezslav Orel. 2001. *Genetic prehistory in selective breeding: A prelude to Mendel.* Oxford: Oxford University Press.

Zohary, Daniel. 1984. Modes of evolution in plants under domestication. In *Plant biosystematics,* edited by W. F. Grant, 579–586. Toronto: Academic Press.

Zohary, Daniel, and Maria Hopf. 2000. *Domestication of plants in the Old World.* 3rd edition. Oxford: Oxford University Press.

Zohary, Daniel, and Pinchas Spiegel-Roy. 1975. Beginnings of fruit growing in the Old World. *Science* 187 (4174): 319–327.

Zohary, Daniel, Eitan Tchernov, and Liora Kolska Horwitz. 1998. The role of unconscious selection in the domestication of sheep and goats. *Journal of Zoology* 245: 129–135.

Agriculture or Architecture? The Beginnings of Domestication

*Peter J. Wilson**

However it is defined, the domestication of humans, plants and animals is a historical phenomenon. It occurred at specific points in time in different regions of the world. It seems as if it first occurred in the Near East about 13,000 years ago. Archaeologists have tended to overlook the spatial and temporal factors of domestication, particularly the use of permanent construction materials, in favor of the beginnings of agriculture. This neglect is particularly important in view of the fact that a built living environment universally precedes agriculture by between 500 and 2,500 years depending on sites and regions. It has hitherto been assumed that building with permanent materials was a consequence of the agricultural needs. Likewise, discussions of contemporary issues raised by domestication, such as many of those discussed in this volume, take for granted the prior necessity to construct a specific spatial environment that will help to produce domestic, institutional, occupational, laboratory, and landscape arrangements and sustain them over time. Often it is the rearrangement of space or the appropriation of place that is at the heart of both intended and unintended consequences, many of which are discussed in this volume.

As domestication has a distinct origin, and as it has universal components and constraints, I outline how, in the epi-Paleolithic and the early Neolithic, architecture not only precedes agriculture but provides the technology necessary for agriculture to emerge.[1] At the same time, I argue that the necessary dimensions of domestication—the appropriation of space, the reconstruction of place, and the marking of time by using permanent materials for dwellings in particular—are basic assumptions that need to be recognized in any discussion of domestication.

Underlying Thoughts

The great Australian archaeologist, Gordon Childe, set the agenda for modernist archaeology (Childe 1952). Modern times, he argued, were ushered in by the *Neolithic Revolution,* which was essentially an economic revolution. This revolution is now conceived as a continuous transformation. Nevertheless, the economic base remains central to theories of prehistoric growth and change. Childe, like others who have followed him, thought of this as a revolution for the benefit of humanity. It laid the foundation for complex societies, and "progress." The key to this was the storage of animals and plants and hence their availability in ever-greater quantities. It is taken for granted this was achieved with the aid of ground and polished stone tools (i.e., neolithic). Jacques Cauvin has recently written that the "Natufian people had no new tool to invent in order to pursue their strategy of farming production" (Cauvin 2000). But hand tools are not the only forms of technology. Was there another technology, hitherto unrecognized?

We should also consider the timing of the introduction of agriculture and animal breeding to the Near East, the region Childe was discussing. This was around 13,000 years ago, probably in *Abu Hureyra,* an epi-Paleolithic settlement on the Euphrates.[2] Gordon Hillman suggests this may have been a reaction to stress—a decline of wild cereal grasses in response to Younger Dryas aridity (Hillman et al. 2001). Although Hillman clearly considers these hunter-gatherers were forced into cultivation climatic circumstances he also writes,

> as hunter-gatherers, their susceptibility to the climatically induced changes in their wild food resources would also have been exacerbated by the fact that they occupied the site *year round* and appear to have done so ever since the site was first settled—11500 14C BP. (Hillman et al. 2001: 390).

These people were sedentary and were so before they grew rye. Ofer Bar-Yosef states that "the establishment of sedentary Natufian hamlets in the Levant marked a major organizational departure from old ways of life" (1998). He does not elaborate, but the burden of his article is that Natufian culture of the Levant was the threshold to the origins of agriculture (Bar-Yosef 1998: 159–177). In the case of *Abu Hureyra,* cultivation of rye began about 500 years after the settlement was established. In the case of the Natufian culture, people appear to have been living in semisubterranean dwellings for nearly 2,000 years before they began cultivating crops. These dwellings were found at *Am Mallaha* (Bar-Yosef 1998: 162). By 10,500 years ago there were several established settlements in the Fertile Crescent, such as *Jericho, Beidha, Netiv Haqdud,* and *Khiami* in the southern arm and *Mureybit, Abu Hureyra, El Kowm,* and *Bougras* in the northern arm (Cauvin 2000). Dwellings at *Am Mallaha* had foundations built of stone supporting an upper structure of bush or wood. At *Mureybit,* they were circular or rectangular (*Jericho*) and built of mud brick over a wooden framework, or of suitably dressed stone. In one instance, *Hayonin Cave,* a series of rooms had been built of undressed stone, with a hearth in each room. One room contained a kiln for burning limestone but was later remodeled as a bone tool making room (Bar-Yosef 1998: 163). For nearly 2,000 years or more people living in these permanent shelters continued to hunt and gather while living in permanently constructed dwellings. Such a life seems to be confirmed by the fact that there were few, if any, storage bins. Toward the latter stages some people of this culture cultivated while others gathered (Kenyon 1957). Finally, by 8,500 B.C.E. they all farmed. The climate was warm, a period known as the *Young Dryas*. The area was relatively small with varying altitudes providing a diversity of climate and fertility, rather like modern California (Bar-Yosef 2001).

If architecture preceded agriculture, rather than the other way round, as is commonly assumed, perhaps we should start to rethink the early *Neolithic*. Obviously the term "neolithic" is too embedded in archaeology for it to be changed. Nevertheless, rather than an improvement in stone tool finishing, the real divide between the Paleolithic and the early Neolithic was the advent of architecture (from the Greek *arche* meaning first, and *tecton* meaning worker). In this chapter, I argue that without architecture there can be no domestication.

Building the Neolithic

The *Oxford English Dictionary* defines *architecture* as the "science of building." *Science* is defined as "systematic and formulated knowledge." In contrast to the temporary shelters, whose remains have been found in the epi-Paleolithic at, for example *Ohalo II* in the Jordan Valley, Israel (Nadel and Werker 1999), the remains of what were once solid buildings are evidence of systematic knowledge. Its basics are identical with contemporary architecture. Archaeologists, when they find remains and ruins, try to reconstruct the same knowledge, working backwards. The archaeologist ends up with the ground plan of a building, which is what the builder began with. For both, what is entailed is the systematic thought and skill that gave shape to the structure in the first place. In contrast, earlier Paleolithic people appear to have had no use for this knowledge. The problems it solved could never even occur to those who moved around and lived for short periods in lean-tos, camps, and caves. There is little in the archaeology of a Paleolithic cave site to reveal any planning.[3]

The earliest buildings of permanent materials—mud, dung, and wood or stone—make a mark on the landscape, they are the first human-made landmarks. How are these landmarks made? First, a site has to be cleared, which creates a shape, that is, it is measured to preplan the structure(s) to be placed on it. We call this *surveying*, based on tri-angulation, and the Greeks knew this procedure as geometry. At the minimum, a clearing must be marked where intended structures are to be placed—the landscape is altered by design.

Whatever the material used, if the structure is to last for some time, as it is completed on site it possesses more and more weight. In con-trast, temporary shelters of leaves and branches have little accumulated weight. We know weight as the force exerted on an object by gravity's pull. The construction of early Neolithic dwellings, using posts to sup-port the walls and roof, suggests the people understood weight as a pressure or flow to be distributed. Epi-Paleolithic and early Neolithic houses were usually semisubterranean, providing a stable, level ledge to support the superstructure. Later, some houses were built on stilts or platforms, which showed a greater understanding of the science of foundations, ground movement, and weight distribution.

Any material has a certain tensile strength such that when subject to stress or force it will fracture or buckle when the force and the ten-sile strength equal zero. We know how to calculate this in advance to prevent fracture or collapse. The structural design, such as the placement of posts and beams suggests early Neolithic people had some idea of this. A mud wall of a certain width can only be built to a certain height and length, unless a series of uprights, posts, or columns are set to a depth into the ground as foundations. These support the wall and evenly distribute the flow of pressure from the roof via beams to the ground. In semisubterranean buildings, the site that was originally cleared must be tamped firm for support and to counter earth movement and water seepage. These might cause shear. Early Neolithic houses in China and Korea, for example, were all semisubterranean and had densely tamped floors (Chang 1986). With shear, one wall, for example, will come away from another wall or cracks will appear, resulting, finally, in collapse. Pressures also originate from inside a building—people moving around, the presence of solid objects and inside warmth or cold all contribute to "within planar pressures" that can weaken a building against the "without planar pressures" of wind, rain, and the more obscure pressures of gravity. Early Neolithic people simply must have possessed this minimum knowledge of structural engineering.

A key feature is the placement of the doorway. This is an empty space, a hole in the wall, where the flow has nowhere to go, so it must be diverted via an arch, most commonly the post and lintel arch.[4] Unless early Neolithic people had systematic knowledge of a practical geometry, which included inventing the simple arch, they would not have been able to build structures that would resist such pressures, as well as the more visible pressures of wind and rain. Furthermore, Neolithic settlements were walled. Like the doorway, the gate, which became the focus of religious processions and transformations, is in an engineering sense its weakest point. It also became a political and sociological tool. Whoever controlled the gate controlled the city.

The Building as Technology

At *Mureybit III* were found the earliest dwellings with compartments. These were stone structures containing roof beams; internal walls; stone platforms, possibly for sleeping; and wooden tool cupboards. The hearth was probably outside. Consequently occupants would be "directed" one way rather than another if they wish to do one thing rather than another—to sleep in one place, to eat in another place, or to keep and fetch tools in yet another place. Internal walls interfere with the direct line of vision allowing some degree of privacy. The door itself creates an orientation for the house as it defines the front or back or sides vis-à-vis other houses, or the settlement wall, or some natural feature such as the sea. Occupants are oriented toward or away from the outside world, for this is where they enter and exit.[5] Modern ethnography provides many examples of the relation between the door and status. Even the simplest dwelling that contains permanent materials in its layout, creates certain distinctions.

The Settlement as Technology

People living within the walls tend to walk a path from their house to the gate when wishing to exit, and from the gate to the house after entering. This suggests a nascent form of routine quite foreign to hunter-gatherers, who follow animal tracks, which are highly variable, or who make paths, which are mostly hidden. Because people must go through the gate they can be watched, protected, counted, or barred from entering or leaving. The gate in the wall is the site of political and religious control, the source of metaphorical exits and entrances. Later in the Near East, the names of cities like Babylon or Memphis were no more than the vernacular terms for "gate." The processions of the gods or kings went from one gate to another, changing identity as they left one and entered another after a journey (3 km in the case of *Marduk* of Babylon). Passing through the gate effects additional transformations to passing over the threshold.

Stated in general terms, the establishment of the buildings made of permanent materials led to ordinary behavior becoming routinized and the door or gateway in particular had a formative influence on sociopolitical control. As soon as there were dwellings in the Near Eastern epi-Paleolithic and early Neolithic there were burials inside. But larger buildings were constructed at the same time. These are called "sanctuaries" by archaeologists. Here, skulls were stored above ground and bones beneath. These structures suggest new routines. One such sanctuary, at *Nevali Çori,* is notably larger than surrounding dwellings, and has a large stone slab inset on the floor, which Jacques Cauvin suggests was a cult building, and another at Cayönü with a red stain, possibly blood (2000: 87). There is also a standing stone that might have been an altar. These, though, are plausible guesses. What we can say with broader generality and more certainty is that such a building, whatever its use, was a focus of public attention at some times. The buildings were large enough to contain an audience, or an assembly, or a congregation. Without specifying which, we can at least say that people gathered there and that all of them paid attention to the same thing at one time. Assemblies or crowds are by no means outside hunter-gatherer practice—Durkheim based his theory of religion on just such a phenomenon—but, in the early Neolithic we are looking at something like community meetings focused on a building that may have been dedicated for a special purpose.

Architecture as Technology

As well as skilled techniques for stone-tool making, engraving, and painting, their technology may well have enabled Paleolithic people to dig and work the ground. Thomas Wynn suggested that the Paleolithic stone ax carried with it evidence of knowledge that had to kept in mind (Wynn 1989). But there appears to have been nothing in the evidence of the remains of their material culture that would suggest full-time, regular working of the soil. However, with architecture there comes the technology that, once its possibilities are recognized, can lead to new developments and consequences.

Lewis Mumford characterized the Neolithic as "the age of the container" (1961). He also suggested that it was a feminine, passive age. I think that architecture can also be considered as an active technology. To understand architecture as technology—a tool to create something new—I would like to return to the coming of agriculture, the domestication of plants and animals.[6] The accepted dogma is that architecture developed subsequent to agriculture because of the need for storage of domesticated crops and animals.

The rhetorical question is: how did cultivators acquire the know-how to build granaries and houses by cultivating? How could they cultivate plants and domesticate animals without having a means to do so? Significantly, there is also the time factor. In the Near East, and in fact everywhere else where domestication was indigenous rather than imported, architecture preceded cultivation. People domesticated themselves first. They made themselves at home.

Storage technology must have been present before people domesticated animals and plants. Not only storage (containment) but also enclosure, for, if you have animals or plants suitably separated by a border, a fence, a ditch, a corral, or a byre, then you have the means, or technology, to control their breeding. It does not matter how much you know about hybridization or crossbreeding; to preserve and continue what you have done, some sort of barrier is the absolute prerequisite. Experiments, or lucky findings, have to be repeated if they are to be successful. In which case there must be a special place where what has been achieved can be preserved. The technology of protection and separation associated with the domestication of plants and animals has become problematic again with the difficulties of keeping genetically engineered (GE) crops separate from non-GE crops—a new form of domestication.

Architecture is composite and one of its main components is the wall. The wall as a separate unit or tool was probably present in the Upper Paleolithic as part of a hunting trap. Two (woven or wooden) fences were set up in a V shape. Animals were driven in and a third fence, a gate, closed the triangle (Russell 1988: 33). The wall was also a part of the temporary shelters built by hunter-gatherers. When these walls were made of fixed, long-lasting materials they clearly suggested new possibilities, although not necessarily immediately. By what we could call "lateral thinking," the dwelling was reconfigured as a different but related construction—as pens for animals or patches for planting. Once they had these tools Neolithic people could domesticate animals and plants. Animal and plant domestication was, technically, a consequence of human domestication. So, as well as the symbolic, aesthetic, and utilitarian properties of architecture it has what is referred to as "agency."

Domesticating the Dead

The act of settlement raises problems of disposing of the dead. Some Upper Paleolithic people did bury their dead and these burials, often quite elaborate, have been taken to indicate religious beliefs, or a belief in an afterlife. At the sites of *Dolne Vestonice* and *Předmosti* graves were found to be beneath the "floors" of the mammoth bone tents—a triple burial in the former site and 24 individuals in the latter. Some individuals were laid on their right side and were flexed, and objects, including figurines, were found (Butler and Hobbs 1996). It is agreed by archaeologists that such sites as these stood for many years and were used for part of the year on a regular basis by hunter-gatherers. Such evidence suggests to me that when people erect structures of durable materials to live in they will also bury their dead there, or nearby. So, in a sense, there was nothing new in the early Neolithic idea and practice in the Near East of burying the dead in or right by the house. It is just that the year-round use of permanent dwellings and their increasing frequency converted what was practiced very seldom earlier into a regular procedure, or custom. China was slightly different. There, cemeteries were created within the confines of the settlement. Early Neolithic graves dug within the dwelling had to have strengthened sides to resist underground pressures and the weight of people walking on them, so they were "built" not just "dug." Overall the contrast between the Paleolithic and the early Neolithic is that in the early Neolithic a subterranean architecture for the dead was created and developed. Graves were originally part of the house, but then, as I have noted, in the Near East special buildings, "sanctuaries," appeared along with secondary burials. These sanctuaries were larger than dwellings so that what began as a subterranean structure was taken to the surface, and then enlarged. If we were to follow the development of tombs and mausoleums through the Neolithic, we would find they come to totally dominate the landscape. The pyramids of Egypt or ziggurats of Assyria and Sumeria and the New World are the obvious examples. In

China, however, larger buildings to house the dead were built below ground. Underground tombs, built as palaces, were of monumental proportions (Chang 1986).

Initially the disposal of the dead might be considered a hygiene problem. But the manner in which corpses were disposed of and the permanent presence of their graves suggests "ontological" thinking, that is, questions of "being." When tombs become monumental and when buildings are highly decorated, they illustrate what the builders, and those who commissioned them, thought of themselves as beings in the future. From the sanctuary at *Cayönü* onward, that is from 9,000 B.C.E., architecture is not only the vehicle but also the place where life, or "being" is concentrated.

Occupation

When people, plants, or animals are enclosed they may be described as occupying space at a particular place at a certain time. In this sense, whenever hunter-gatherers camp for days and nights at a good spot they occupy that spot. When they leave, the only signs of their occupancy are the remains of a fire and some detritus. In desert country, this may be obliterated in a few days. When people occupy caves the reminders of their sojourn last longer, but their length of occupancy is indeterminate. Sometimes the cave itself was modified and lived in for long periods of time. When people build habitats of permanent materials they become occupants. There is a change of status and identity—all the occupants of a habitat have something in common, which distinguishes them from occupants of similar but separate habitats. From early Paleolithic sites we know there were often specific places where stone tools were made— usually near the quarry. From the epi-Paleolithic and early Neolithic this pattern is continued, but with a significant change. Workplaces are located in, or beneath, or close by houses. The constructed living site becomes a place that fosters an occupation—a trade or a skill. Donald Sanders (1990) provides a wonderful analysis of the use of space in a domestic setting using data from his work at the Bronze Age site of *Myrtos* in southeastern Crete. He notes there were places for work, for sleeping, for cooking, and that there were public and private places. He mainly infers all this by working out the lines of vision that occupants would have had from various rooms. It is clear that different occupations were recognized (Sanders 1990).

Different occupations take up different places. In Babylon, streets of the "upper class" were lined with two storied buildings. These contrasted with the small two roomed huts of the "lower classes"—if indeed there were classes as we understand them. Wealth does not always define segregated classes of people, or people of a different kind. However, architecture always does. The peasants of the Pharaohs' Egypt lived in small mud-brick houses clustered in villages that were dwarfed by the palaces, temples, and tombs of aristocrats and high executives. The progress of architecture not only fueled, but was fueled by the cultural differentiation and grouping of people from the most immediate level, that is, those occupying the house, to the widest and highest level of the state. Put simply, architecture from at least the early Neolithic was used to put people "in their place," which they could occupy themselves and work at their occupation.

The chapter in this volume by Feeley-Harnik is not only a fascinating slant on Darwin's insights but it is also a prime illustration of the link between domestication and occupation. The "occupation" of silk weaving "occupied" one area of London, Spitalfields. The streets and houses of Spitalfields were places where the "occupation" of pigeon fancier was concentrated and professional, even though it was a hobby. Then, when mechanization replaced the skilled hands of the weavers they had to move out, and Spitalfields was taken over, or "occupied," by others.

Institutions

The thoughts and deeds I have mentioned have in common an association with place. More than that, they are associated with human beings making places and then dwelling in them as occupants or inhabitants. When this happens, what were once memories, ideas, and beliefs that previously could only have been communicated orally or mimetically are set out by the construction of a place. With domestication and such things as "house and settlement rules," some activities become organized and arranged. They become what I call *institutions*—a prime result of architecture as "agency."

In our present thinking, institutions refer to rather well-established real or imaginary assemblies of people who are bound together by a common interest or occupation governed by rules administered from a headquarters—a special building. For example, "the family"; Parliament or Congress; No. 10 Downing Street or the White House. Organized, regular activity involving several individuals, living or dead, whether it be a kula voyage, a rain dance, a communal meal, a parade honoring the dead, cooking in a reserved part of a building, a funeral or a dance, a lecture or an examination, a conference, or a shareholders meeting are even more realistically "institutions." They are vital to the survival of the community and its culture. An integral feature of their organization is the design of the place where they are conducted. Frequently a structure is purpose built as part of the formation of the institution—the college, the ritual corners of a Trobriand garden, the Neolithic sanctuary. The placement of activities—consider this conference or your university—is essential to the design and purpose of that activity. Universities, hospitals, libraries, laboratories, national parks, wilderness areas, factories, offices, bars, airports, churches, and apartments are places designed to help guide and organize a set of activities that individuals may perform in unison or individually but in a patterned or constrained manner. Hierarchy, for example, may be marked in a U.S. corporation's building by floor height.

The English word institution derives from the Latin *instituto,* which means design. Once places are built, whether they are open dance circles cleared of trees and scrub or buildings raised above the ground using processed natural materials, they impose a design. Even the designation of a region as a "reserve" is accompanied by various forms of marking out such as notices, entrances, fences, and uniformed management personnel (see, e.g., Suzuki this volume). In the early Neolithic, this designing of the landscape and buildings was led to the first larger-than-human constructions. Such designs come between people and their natural environment. They have to negotiate them and they organize elementary features of behavior like taking a route, assembling together, or trespassing. Architectural design of and on the landscape is the tool that permits people to form institutions. Whereas people without designed places may certainly have many thoughts about the world they live in and the activities they undertake, they tend to lack regularity and organization. But once people have made a place for themselves, their ideas and behavior have a framework. This framework persists through time. The most common framework is the house, then the settlement and the land centered around the burial place. Symbols only have meaning when they are in place over time.

The creation of place by architecture in the early Neolithic together with the proliferation of relics and remains of objects such as figurines, statuettes of numerous subjects including animals and people, ornaments, and, later in the Near East, of pottery, is a tangible hint of the rise of institutions. To Jaques Cauvin it suggested the beginning of religion, which sparked a "psycho-cultural" revolution, implying that agriculture was driven by religious rather than economic needs. I am not suggesting that prior to the use of architecture and agriculture, behavior was mindless, chaotic, and disorganized. Hunter-gatherer ethnography supplies plenty of evidence to the contrary. But, comparatively speaking, foragers lack the "architecture of the institution." They are, as their ethnographers indicate, flexible in their associations and fluid in their organization.[7] There is a parallel between what disappears after a modern hunter group quits camp and the same lack of evidence of organized living to be found in the Paleolithic. There are exceptions, such as the remains of mammoth bone buildings dating back to 20,000 B.C.E. at *Kostenki, Dolni Vestonice,* and other sites then at the verge of the permafrost regions in Siberia and Russia (Soffer 1985). Or the oldest camp site of all, *Terra Amata,* going back possibly more than 400,000 years. But climatic conditions were severely limiting and there was no possibility of domestication. It is this comparative lack of variety and rigidity as well as quantity of evidence that points to the increasing divide between the Paleolithic and the Neolithic. It is only when people established places for activity that we have a better idea of what they were up to and can appreciate the beginnings, and then development of the institutionalization of behavior. Institutions, distinct ways of doing things in particular places, are the realities we live with, and in, rather than "imagined" societies and cultures.

Building Thoughts

Colin Renfrew has spoken of the "sapient behaviour paradox," by which he meant that there is such evidence in the Upper Paleolithic as the wonderful cave paintings of France and Spain, beautiful stone work, and engraving and carving of bone (Renfrew 1996). Yet we do not get any sense of a development of these skills ramifying into different aspects of life throughout the Paleolithic. Then, in the short period of 13,000 years, we go from stone tools to nuclear reactors, from tally marks to computers. His explanation for this is sedentism, but he does not elaborate. He also suggests, in another work, the missing cognitive link—the development of "constitutive symbols" (Renfrew 2001). People experience something like the different heaviness of various rocks when they pick them up and then formulate the general concept of weight from comparison. This, Renfrew suggests, allows them to think better and more inclusively. But people have different weightlifting abilities, so although one individual may be able to form a general idea of one thing being heavier than another, it cannot be a general concept. That must be an "independent" standard of measurement, and there is no evidence of this until later in the Neolithic. The beginning of architecture, though, at least provides "concrete" (or mud or stone) evidence that illustrates such general concepts.

To repeat, a structure built of permanent materials has components—it is a composite tool. There are upright posts and horizontal struts while placing beams and posts creates angles and differences of height, width, weight, and other factors that have already been discussed. Irrespective of each of these features as a part, the whole they compose is on a scale the same size as, or larger than, the person. It has to be put together by men and women handling materials and tools, not just in a certain order but in particular ways to create the whole structure. The sum is certainly more than the parts. To get the angles of support right to prevent fracture, to get the correct relation between height and width to forestall collapse, and to construct the door to the proper proportions to prevent cave-in, general concepts are necessary and implicit (or even explicit) in the structures. Furthermore, no matter how simple the building, when putting up one part account must be taken of what has already been erected and what is yet to be erected. If a building is not to collapse, it has to be built to plan and measurement. There are also the constraints of the materials used and the natural forces that are brought into play have to be foreseen as building proceeds, and countered before they occur. This depends not only on good workmanship but also on good design. Once you have a good basic design you also have a plan for future structures—a building is a model of, as well as a model for. Although they may not have been the earliest, the general concepts of measurement that permit constructing with permanent materials are the first evidence we have for general concepts.

I am writing only of single structures, but in the early Neolithic period of the world's civilizations buildings usually occur in settlements, often only small to begin with, but settlements nevertheless. And in the Near East these were often walled. So within a settlement we have an arrangement of designs—the dwellings and the spaces between them. Irrespective of pattern, whether settlements are circular, oval, rectangular, or whatever, structures have some sort of orientation, and practical arrangements have to be made. Disposal places, for example, are not an issue for hunter-gatherers, but they are for domesticated people. These are multiplied when they keep animals. So, without going beyond the simple facts of living, architecture compels people to think in a patterned manner, fitting together many considerations at one time. Either one person has to think of several things at the same time, or several people have to think of the same thing at one, or different, times. And they all have to fit together. This does not necessarily mean that Neolithic thought replaced Paleolithic thinking, that there was an evolution of thought, but, rather, the one incorporated the other.

Architecture and Creativity

In her classic book *The Art of Memory,* Frances Yates gives an account of Cicero's system for remembering long speeches (Yates 1966). You look around the hall in which your speech is to be delivered and you attach successive themes to features of the hall. Later you "read off" your speech from the hall. This system was on sale as recently as the 1970s.[8] Many preliterate cultures go further than this and use the house as mnemonic for the cosmos or for recording their ancestral history. The house is often the "computer" that does the "equations" between the human body, the physical environment, and the cosmos. Ethnographies from Amazonia, West Africa, and Southeast Asia in particular provide vivid examples of this.[9] Until the invention of writing, possibly in the third millennium B.C.E., although maybe much earlier in China, buildings were what Merlin Donald (1991) called *External Storage Systems,* although he does not realize it. One cannot claim that modern meanings of the house were the same for early Neolithic people. All I want to suggest is that architecture itself can hint at connections that were no part of its construction. And once recognized they may be put to use—as is clear when people built "houses" for the gods, for the dead, for rulers, and offices for priests and functionaries.

Buildings provide "cultural" shapes and the visible parts of their structure, such as roof beams, reveal an unnatural, regimented, repetitious geometry. Until recently, battles were geometrically organized the better to exercise force strategically and tactically. Ceremonial military parades and maneuvers are geometrical, the better to portray power by display. Ever since the beginning of the era of large-scale building in the middle and later Neolithic, there has been a dialog between the roundness and pliability of the human body, the angularity of buildings and the irregularity of nature. Jacques Cauvin, when he rushes from statuettes to "Gods" and "Goddesses" has no right to do so. All we can really say is that they are "fat ladies" and "muscled animals" (bulls and lions). Rather than, or maybe as well as, being religious they could be for aesthetic enjoyment. The Greeks attempted to reconcile this contrast between the body and architecture by using the human body as the proportional measure for their columns and modules (Vitruvius 1999).

Amos Rapoport describes architecture as having "low criticality" by which he means a place can be used for many activities even though some places are reserved for particular uses (1990). It can also act in quite different ways at the same time. A building can be a dwelling, a representation of the human body, a representation of the cosmos, a model for building, a container for plants and animals, or for the dead— all at the same time. Living in this immediate surround of possibility was to become an everyday experience for most Neolithic people. As the archaeological record of the development of the Neolithic shows, domestic living became a multimeaning experience and, whether in the making or in the dwelling, architecture was the physical and intellectual core of a system of living—domestication.

For the first time in the history of human thought, early Neolithic people experienced structure, as well as built it. For the first time people built a system that could, and would, become a new source of a powerful metaphor leading on to further understanding, or theorizing. Although we do not know that early Neolithic people used architecture as a metaphor, we do know that architecture changed the world people lived in, that is, for the first time culture changed nature.

Time

Thus far, I have attended to place but not time. The sense of time as an "arrow" or as a "circle" appears absent from hunter-gatherer societies.[10] There is always the sequence of night and day, and in subtropical and temperate zones, time and seasons perhaps were, and are, closely associated. The body too has its changes and rhythms. We should also bear in mind the degree of astronomical knowledge possessed by people living different ways of life. Any such prehistoric knowledge can only be supposed. It is only when building begins that there is solid evidence of the concept of time, of the marking of time, often associated with astronomy and most famously with the henges of England.

There are two features of the early Neolithic that are relevant to the invention of the concept of time. First, there is the burial of the dead inside the house. There is evidence from the sites mentioned earlier of burials on top of one another. This suggests a "household" passage of time, a sort of history—either of lineage or of occupancy, or both. There is, however, no evidence this was recognized as such. Although buildings were made of permanent materials, the term "permanent" is relative to the human lifespan. Consistently from the earliest settlements, and all over the world, new houses and entire settlements are frequently built on top of ones that have gone to ruin. Sometimes this rebuilding was continuous; sometimes sites were abandoned and then recommissioned. Either way these reconstructions are possible cultural *records* of past human activity. They incidentally mark the passage of time of "people without a history." However, once buildings of monumental scale appear, history materializes, and the buildings become markers. Often, single architectural features such as pillars and columns are the architectural marker, although in the present day the commemorative wall has found favor. Statues and busts of kings and queens or other notables, especially when they are portraits, are recognition that there will be a history. Writing of Van Dyck's famous portrait of Charles I, Ernst Gombrich says it

> showed the Stuart monarch as he *would have wished to live in history*: a figure of matchless elegance ... a man who needs no outward trappings of power to enhance his natural dignity. (Gombrich 1972: 316–317; emphasis mine).

This may not be true of early Neolithic figurines and statuettes, but once we come to the age of Assyrian depictions of kings in battle or the friezes of the great buildings of Ur or, later, the statues, busts, and mummies of Egyptian royalty and nobles, this is surely the case.

I am suggesting that with the beginning of burial in the house, or the "domestication of the dead," time and space are brought together in human thinking for the first time. At least, for the first time, we are provided with material evidence of this. Other early Neolithic appearances signifying time also occur. There is, for example, evidence of bones used for scapulamancy in the early Neolithic of China. Scapulamancy can be used for several things such as diagnosing illness or giving advice. I admit we cannot say exactly what the early Neolithic Chinese people used these bones for, but among the possibilities is prediction. When we come to the preservation and specialized storage of the dead in magnificent buildings that dominate the place—that are the place—time and space are the undoubted external dimensions of thought, Cartesian "extensions." Writing, often mistaken as the representation of speech, is an external storage and enlargement of memory, which is time. Possible evidence of the earliest Chinese writing has recently been found engraved on a tortoise shell dated 6,000–7,000 years ago (Li et al. 2003). Prototypic writing appears in the geometric decoration of pottery, which originated far later than building in the Near East but much earlier in the Russian Far East.

Early Neolithic Patterns of Thought

Of course there are problems with the claims I am making for the domestication of human thought. How does a supposedly "universal" human brain "suddenly" enable itself to think in terms of patterns, designs, arrangements, and plans when for millennia—since the very beginning—it only needed to deal with the world more straightforwardly and in its own, opportunist and isolated terms? My answer is that this became possible whenever and wherever the people of the early Neolithic substituted permanent materials for temporary ones when building themselves shelters. Relationships between the person and the environment, and among people, were realigned in the sense that whereas the hunter-gatherer (from ethnographic evidence) views the environment as a provider and protector, the builder extracts from the earth itself. And then imposes a construction. From the epi-Paleolithic on, humanity has extracted from the earth and increasingly covered it with its own, architectural, creations. The mode of thinking about the world has changed from "identity with" to "distance from."

If this notion, together with the evidence I have offered, is accepted, then, eventually, research leads anthropology to join cognitive psychology. What we have to think about is the other beings and objects of the world. Our emotional intelligence guides our actions and attitudes and our cortical intelligence searches for patterns leading to understanding, not least of our emotions. Sensory deprivation studies demonstrate how the brain not only goes to work but also develops in an environment, especially at crucial trigger times. From a cognitive standpoint the people of the early Neolithic inaugurated the production of the built environment that now overwhelms nature and us. There are more objects to relate to and understand than were ever dreamed of in Paleolithic times. As Gilbert Ryle wrote 50 years ago:

> When we describe a performance as intelligent, this does not entail the double operation of considering and executing. ... Overt intelligent performances are not clues to the workings of the mind; they are those workings. (Ryle 1949: 30, 52)

What the archaeologist and anthropologist does is study the results of past and present intelligent performances, performances that have been created as products of, and stimuli for human brains to continue to work. A watershed in human thought and performance occurred between 18,000 and 13,000 years ago. It was the construction of a designed environment, or architecture that began a change in human living that led to agriculture. From this there was no going back to the way of life human beings had led for millions of years. In a word, they became domesticated. This suggests a more general proposition: if domestication is to occur then the landscape must be reconstructed.

Notes

Editor's Note: Sadly, Peter died before he was able to revise his chapter in the light of comments from readers. Despite this, we have chosen to reproduce it here because it was an important catalyst for discussion in Tucson, and Peter himself contributed greatly to the success of our symposium. We hope that it will be read as a suggestive piece, which highlights the importance of material culture, and the spaces in which domestication takes place. As such, it provides a contrast to the other chapters' focus on relationships between living things in particular historical and ethnographic contexts.

1. Although the term *architecture* tends to conjure up ideas of monumental buildings, it refers to the design and construction of any and every humanly built environment. I am aware that bee hives and ants nests are often described as architecture, but this is a metaphorical usage.

2. This is the modern date and not the date used by Childe.

3. Some caves were modified. A protective ledge may have been built, hearths were constructed, and, in some cases, people were buried in caves.

4. Or the roof. In the remarkable site of the former city of Çatalhöyök, in Anatolia, entry was by a trap door in the roof, similar to the manner of entry to the Pueblos of the U.S. Southwest. Helen Leach ventured a suggestion that ladders may have originated in the Paleolithic to reach tall tree fruits. However, contemporary hunter-gatherers usually use ropes for climbing trees.

5. Orientation of temporary huts was used by the Mbuti as a means of expressing disapproval (or approval) of neighbors (Turnbull 1962a, 1962b). The same practice was used by the Naiken of the Nilgiris in South India (Bird 1983). Like much else, the use of the doorway became "fixed" once construction was with permanent materials.

6. *Webster's* defines *tool* "as an instrument of manual activity, particularly as used by farmers and mechanics." The *Oxford English Dictionary* gives a figurative meaning "things used in an occupation or pursuit."

7. The terms "fluid," "flexible," and "flux" used to describe hunter-gatherer ways of life were introduced by Colin Turnbull in 1962b.

8. The system was known as the Pelman System of Memory and was advertised by promising that you would "amaze your friends with your powerful memory."

9. Some examples are Roxanne Waterson (1990), Goldman (2004), and Griaule and Dieterlen (1965). On a worldwide scale, Enrico Guidoni's wonderfully illustrated book is still unbeatable (1978).

10. Although the still-controversial inferences made by Alexander Marshack about Paleolithic lunar calendars and tally marks should certainly be borne in mind (1972).

References

Bar-Yosef, Ofer. 1998. The Natufian culture in the Levant, threshold to the origins of agriculture. *Evolutionary Anthropology* 6 (5): 159–177.

——. 2001. From sedentary foragers to village hierarchies: The emergence of social institutions. In *Origin of human social institutions,* edited by W. G. Runciman. London: British Academy.

Bird, Nurit. 1983. Conjugal units and single persons: An analysis of the social system of the Naiken of the Nilgiris (South India). Ph.D. dissertation, Cambridge University.

Butler, Ann B., and William Hodos. 1996. *Comparative vertebrate neuroanatomy: Evolution and adaptation.* New York: Wiley-Liss.

Cauvin, Jaques. 2000. *The birth of the gods: The origins of agriculture in the Near East,* translated by Trevor Watkins. Cambridge: Cambridge University Press.

Chang, Kwang-Chih. 1986. *The archaeology of China.* 4th Edition. New Haven, CT: Yale University Press.

Childe, Vere Gordon. 1952. *New light on the most ancient East.* 4th edition. London: Routledge and Kegan Paul.

Donald, Merlin. 1991. *The origins of the modern mind.* Cambridge, MA: Harvard University Press.

Goldman, Irving. 2004. *Cubeo Hehénewa religious thought: Metaphysics of a Northwest Amazonian tribe,* edited by Peter J. Wilson. New York: Columbia University Press.

Gombrich, Ernst. 1972. *The story of art.* 12th edition. Oxford: Phaidon Press.

Griaule, Marcel, and Germaine Dieterlen. 1965. *Le Renard pale.* Paris: Institut d'Ethnologie.

Guidoni, Enrico. 1978. *Primitive architecture.* New York: Harry N. Abrams.

Hillman, Gordon, Robert Hedges, Andrew Moore, Susan Colledge, and Paul Pettit. 2001. New evidence of lateglacial cereal cultivation at Abu Hureyra on the Euphrates. *The Holocene* 1 (4): 383–393.

Kenyon, Kathleen. 1957. *Digging up Jericho*. London: Ernest Benn.

Li, Xueqin, Garman Harbottle, Juzhong Zhang, and Changsui Wang. 2003. The earliest writing? Sign use in the seventh millenium B.C. at Jiahu, Henan Province, China. *Antiquity* 77: 31–44.

Marshack, Alexander. 1972. *The roots of civilization*. New York: McGraw-Hill.

Mumford, Lewis. 1961. *The city in history: Its origins, its transformations and its prospects*. London: Secker and Warburg.

Nadel, Dani, and Ella Werker. 1999. The oldest ever brush hut plant remains from Ohalo II, Jordan Valley, Israel (19,000 B.P.). *Antiquity* 73: 755–764.

Rapoport, Amos. 1990. Systems of activities and systems of settings. In *Domestic architecture and the use of space: An interdisciplinary and cross-cultural study*, edited by Susan Kent. Cambridge: Cambridge University Press.

Renfrew, Colin. 1996. The sapient behaviour paradox: How to test for potential. In *Modeling the early human mind*, edited by Paul Mellars and Kathleen Gibson. Cambridge: McDonald Institute Monograph Series.

——. 2001. Commodification and institution: On group-oriented societies and individualizing societies. In *The origins of human social institutions*, edited by Walter G. Runciman. London: The British Academy.

Russell, Kenneth. 1988. *After Eden: The behavioral ecology of early food production in the Near East and North Africa*. London: BAR International Series 391.

Ryle, Gilbert. 1949. *The concept of mind*. London: Hutchinson.

Sanders, Donald. 1990. Behavioral conventions and archaeology: Methods for the analysis of ancient architecture. In *Domestic architecture and the use of space*, edited by Susan Kent. Cambridge: Cambridge University Press.

Soffer, O. 1985. *The Upper Paleolithic of the Central Russian Plain*. New York: Academic Press.

Turnbull, Colin. 1962a. *The forest people*. Garden City, NY: Doubleday.

——. 1962b. The importance of flux in two hunting societies. In *Man the hunter*, edited by Irven De Vore and Richard Lee. Chicago: Aldine-Atherton.

Vitruvius. 1999. *Ten books on architecture,* translated by Ingrid D. Rowland, with commentary and illustrations by Thomas Nicholas Howe. Cambridge: Cambridge University Press.

Waterson, Roxanne. 1990. *The living house: An anthropology of architecture in South-East Asia.* Singapore: Oxford University Press.

Wynn, Thomas. 1989. *The evolution of spatial competence.* Urbana: University of Illinois Press.

Yates, Frances. 1966. *The art of memory.* London: Routledge and Kegan Paul.

Monkey and Human Interconnections: The Wild, the Captive, and the In-between

Agustin Fuentes

What does it mean to be "wild"? In many cultures, especially across the Western world, the answer frequently lies in looking to other animals, particularly mammals. How humans see and use animals and how humans impact the "nature" of other animals are topics familiar to academic and popular literatures (Corbey and Theunissen 1995; Lévi-Strauss 1963; Mullin 1999; Rothfels 2002). Recently, anthropologists have begun to produce a framework that reenvisions cultural and eco-logical complexities and encourages the perspective of dynamism in our examination of the human–animal interface. Work challenging static and purely functional perspectives on domestication and incorporating the cultural conceptualizations or representations of animals are two main focal points that emerge as pivotal in this dynamic approach (Cassidy this volume; Leach 2003, this volume; Mullin 1999, this vol-ume). It is my contention that these newer perspectives are especially applicable, and highly valuable, when our anthropological lenses turn toward a focus on the interconnections and interactions between humans and the nonhuman primates (Corbey and Theunissen 1995; Fuentes and Wolfe 2002; Haraway 1989; Janson 1952; Ohnuki-Tierney 1987; Paterson and Wallis 2005).

Because of the biological, phylogenetic, and behavioral overlaps be-tween humans and the nonhuman primates, relationships between the two groups have a special significance (Corbey and Theunissen 1995; Fuentes and Wolfe 2002). No other organisms on the planet share as much structurally and behaviorally with humans. Humans, along with other primates, exhibit a series of primatewide trends such as grasping hands, relatively large brains, an emphasis on visual signaling, extended infant dependency, and behaviorally dynamic social complexity. The existence of these primatewide trends in both morphology and behavior suggest the possibility that interconnections between humans and other primate species may differ from those between humans and other mammals. The commonalities in physiology, brain structure, evolutionary histories, and the resulting behavioral and social patterns suggests that humans and other primates may share a significant "moment," or pattern or context of physioemotional connectivity (see Clark this volume), not present in the same manner between humans and other nonhuman animal groups. However, these similarities (esp. the social complexity and manipulative abilities) make nonhuman primates very unlikely candidates for the traditional forms of domestication as with many other mammals, such as bovids, equids, and suids. Because of the limited ways in which humans can benefit from engaging in direct domination and control of primates and the extensive geographical overlap between human and nonhuman primate species the relationships between humans and the other primates are not directly comparable to other "domesticate" mammalian species. However, these human–nonhuman primate similarities increase both the likelihood of cultural association or inclusion of other primates by humans, and certain primates' potential to coexist with humans. In short, some species of primate may fall very directly into the ecology, and culture, created by the human niche.

Emerging from the chapters in this volume is a shift in the way we view domestication, a shift that I interpret as moving the focus away from static definitions focusing on human manipulation and control of other organisms. For me, this shift moves the focus to the contexts and patterns that characterize the interface between humans and the animals they share space with. Some of these relationships might be viewed as "domesticatory practices." Here, defined as human action and alteration of ecologies impacting and shaping the behavior and physiology of the species in and around the humans. This results not in traditional domestic species, but species that are being directly shaped by processes (domesticatory practices) resulting from human action.

A portion of the human adaptive success is effectively represented as niche construction (Fuentes 2004, 2006a; Odling-Smee, Laland, and Feldman 2003; Potts 2004), the creation of human *place* (sensu Richardson 1989). If other primate species are living in or around this human *place* we can attempt to assess its impact on nonhuman primates as a patterned process. As is evident from the rapid and dramatic reduction in numbers and range of some species of primates, many primates do not do well when they encounter human *place*. However, some species, such as macaque monkeys, do quite well around humans. By examining specific contexts of this monkey–nonhuman primate interface we can view a set of scenarios wherein human niche construction and particular human cultural behavior act as domesticatory practices creating ecological patterns and contexts in which certain species of primates can effectively exist. The relationship between humans and the members of the genus *Macaca* (macaque monkeys), especially in and around Asian temples, is an excellent example of this complex relationship (Fuentes, Southern, and Suaryana 2005; Wheately 1999).

Temple Monkeys in Bali

In 1989, I visited the temple forest complex at Padangtegal (Ubud), Bali, Indonesia, for the first time. There I saw macaque monkeys and humans coexisting and interacting on a daily basis with little significant conflict. The macaques used the surrounding forest, the temple structures and the neighboring village gardens and rice fields. The local Balinese used the temples, the paths through the forest, and managed the neighboring gardens and rice fields. Over my years of visiting and researching at the site, I realized that the relationship between both species of primate is complex, with mutual impact. The macaques are protected when in and around the temple site and the humans gain substantial financial benefit from the monkeys' presence through tourism. The macaques also play important symbolic roles in Balinese mythology and recent work suggests they may be important factors in the disease ecology of Bali itself (Engel et al. 2002; Jones-Engel et al. 2005). My subsequent research across the island of Bali demonstrated that the complex socioreligious–agricultural patterns of the Balinese created a space and ecology (Balinese *place*) that favors the macaques over all other large "feral" mammals on the island. At the same time, this also impacts the structure of the macaque populations (Fuentes, Southern, and Suaryana 2005; Lansing 1991). Expanding my research into human–macaque relationships across Asia drove home a particular point: macaques and humans have complex social and ecological relationships in many locales throughout Asia.

This realization led me to examine more broadly the relationships between humans and nonhuman primates. Although most of these

relationships may not fall under the traditional rubric of domestication, I contend that there are patterns and contexts wherein the lives
and ecologies of the primates are significantly shaped by humans. This
molding of many primate species by humans may be unintentional, but
nonetheless might be seen as a domesticatory practice. I suggest both
that primates play important roles in human cultures and that human
cultural behavior (human *place* as our niche) creates ecologies that
differentially impact primates. This chapter is an attempt to examine
various human–nonhuman primate interconnections with a focus on
those best seen as involving domesticatory practices.

Human–Nonhuman Primate Relationships: The Wild, the Captive, and the in-between

Although human and nonhuman primates share a number of interconnections, at the most basic level there is a distinction introduced
by geography. When envisioning the globe (fig. 5.1) there are relatively
clear demarcations between zones of sympatry (geographic overlap)
and zones of allopatry (lack of geographic overlap). These geographic
patterns impact the context and content of interactions between human
and nonhuman primates. The zones of sympatry fall primarily in Africa;
South, East, and Southeast Asia; and South–Central America and reflect
the distribution of nonhuman primates for at least the last 10 millennia.
The zones of allopatry are primarily above ~30° N of the equator (above
the Tropic of Cancer), except in East Asia where two species of macaque
monkeys *(Macaca mulatta* and *M. fuscata)* range relatively far North
in China and Japan, and North America, where nonhuman primate
populations barely reach the tropic of Cancer (23.5° N).

Understanding the relationships between humans and nonhuman
primates requires a temporal context as well as a geographic one. Long-
term sympatry, especially sympatry that involves common usage of habitat
can result in a form of coecology. Ecological pressures impact mammals
in particular ways, and mammals that share many morphological and
physiological facets in common, such as the anthropoid primates (including monkeys and humans) may share similar adaptations. This is
important in understanding the interconnections between humans
and nonhuman primates because long-term overlap and similarity in
behavior and physiology can impact human conceptualizations of
"nature." This also may act to facilitate distinct patterns of integration
and engagement between the humans and other primates, especially
with macaque monkeys (see below). Alternatively, areas with less time

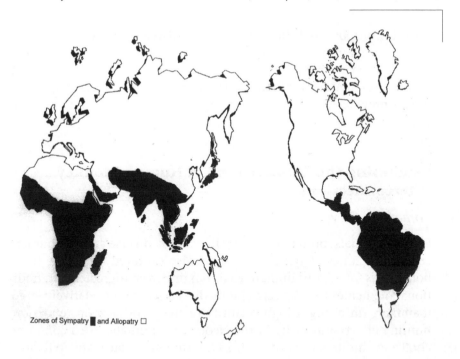

Zones of Sympatry ■ and Allopatry □

Figure 5.1 Zones of Sympatry and Allopatry.

overlap, especially in zones of allopatry where overlap is mitigated by captivity or other forms of limited or selective exposure, specific patterns of strong association and incorporation are expected to be uncommon.

We can characterize human–nonhuman primate interconnections in three broad categories: the wild, the captive, and the in-between. "Wild" interconnections refer to naturally occurring, overlapping populations of both humans and nonhuman primates in the zones of sympatry. The "captive" are interconnections characterized by allopatric situations in which no free-ranging nonhuman primates overlap with human zones. Examples include zoos and safari parks but also exposure via entertainment (circus, cinema, etc.), or literature (popular fiction and academic or informational texts). Here the "in-between" represents the myriad of contexts where there are variations on the themes of wild and captive. This "in-between" label includes those primates who excel in complex human modified habitats, and includes areas where primates are integrated into cultural practice well beyond simple geographic overlap and/or exploitation by humans for food.

Here, I propose that there are four "in-between" relationships of relevance: primates as prey, primates as pets, primates as social and economic participants in human culture, and the special case of macaques at temple sites in Asia. This final relationship can be seen as an integration of the previous three and may most clearly reflect a domesticatory practice.

Examining the "In-between" in Human–Monkey Interconnections

Primates as Prey

Being mammals, primates are prey items for a diverse array of human cultures in zones of sympatry and are also captured by people from both zones for various human needs ranging from ingredients in traditional medicines to subjects of biomedical research. As relatively large mammals (meaning a high quantity of meat and a good return on hunting effort), especially in the Neotropics, primates are a common choice for hunters (Alvard et al. 1997; Amman, Pearce, and Williams 2000; Lizarralde 2002; Shepard 2002). However, there is no evidence that humans take nonhuman primates randomly or solely in respect to optimal prey-return models (Shepard 2002). In fact, it is common for human hunters to selectively hunt specific primates over others for very specific historical, religious, social, taste preferential, and a myriad of other reasons (Cormier 2002, 2003; Fuentes 2002; Lizarralde 2002; Shepard 2002). Although human predation on primates is not clearly a domesticatory practice, it can act to influence the behavior and genetic structure of primate populations by selectively removing individuals from the breeding pool.

In addition to traditional predation (hunting for consumption), some nonhuman primates are currently raised in captivity and selectively bred for human biomedical experimentation. These colonies (primarily of macaque monkeys and a few other species) are selectively bred to create specific pathogen-free breeding colonies to provide the appropriate test subjects for research into a variety of human disorders ranging from obesity to HIV/AIDS. The breeding and sacrificing of specific physiologically marked and altered populations of nonhuman primates does approach the traditional, directly functional definition of domestication.

Primates as Pets

Although "pets" is a substantial category, here I narrowly define it as animals kept for companionship, enjoyment, or status rather than as working "domesticates" that contribute to the income or nutritional intake of the humans who own them. Today it is largely in areas of sympatry that we see primates as pets. However, the present situation is largely recent as nonhuman primates, especially small monkeys, were widely available in Europe during the early centuries following colonial expansion. Both monkeys and apes were kept as pets in private menageries and homes during that time. However, because of the costs of transport and the high levels of mortality during transport it is unlikely that the zones of allopatry ever had the levels of primate pet ownership that we see in zones of sympatry. Today, pet ownership of select nonhuman primate species is widespread in Southeast Asia, and Amazonia, and occurs in South Asia, and other parts of South America and Africa as well.

It is not clear that the keeping of pets impacts the "wild" populations of monkeys and apes in a way distinct from predation, as pets are frequently by-products of hunting, and capture for pet-keeping functions similarly to extraction via predation. Because the pet tradition is quite old in Southeast Asia (and probably South America as well) it is also quite likely that there is substantial bidirectional pathogen transmission between humans and their primate pets (Jones-Engel et al. 2001). It is possible that there are physiological changes, even adaptations, in populations of human and nonhuman primates that overlap extensively, such as pets, humans, and the local populations that the pet primates come from, as a result of these intensive interactions.

Social and/or Economic Participation

Here, I refer to the participation or inclusion of nonhuman primates into human society. Although there are a number of human societies in which this pattern takes place, the Guaja of Brazil are among the strongest examples (Cormier 2002, 2003). The Guaja, a Tupi-Guarani speaking group in Amazonia, display an intense inclusion of nonhuman primates into their social fabric. Young monkeys are kept as pets and trained by children, cared for by women, and represent a specific image of fertility highly valued by the Guaja (Cormier 2002, 2003). Women will bathe, breast-feed, and "wear" the monkeys. Throughout much of their development, the monkeys are carried and cared for by members of the Guaja village (Cormier 2002, 2003). Although pattern of carrying and raising monkeys has a considerable cost in terms of energy, movement, and food stuffs, the returns go beyond the purely social. Young girls frequently practice mothering with infant monkeys and boys gain important hunting experience through exposure to monkey movement and behavioral patterns (Cormier 2002, 2003).

Nonhuman primates are also used as economic tools in South, Southeast, and Northeast Asia. Although there is reference to crop-picking macaques across much of Southern Asia, the best example comes from Thailand where males of one species of macaque *(M. nemestrina)* are kept, raised, and trained to pick coconuts (Sponsel, Ruttanadakul, and Natadecha-Sponsel 2002). In this case humans in Southern Thailand capture macaques as young individuals and then invest substantial time and energy to train them to be effective coconut pickers. This system can be highly efficient for the humans as a well-trained macaque can harvest between 500–1,000 coconuts per day and do so with a lower cost and higher return than a human could (the other harvesters are young human boys). Sponsel, Ruttanadakul, and Natadecha-Sponsel (2002) argue that cultural and agricultural systems have favored a move from conflict between humans and macaques over crops to a synergistic relationship wherein the humans capture, train, and maintain the macaques as they serve their economic role.

Monkey performance is a blend of the economic and social inclusion. Monkey performances include a variety of theatrical performances by trained macaques, in which the monkeys mimic human cultural behavior via a series of staged interactions with their trainer, the audience, and other monkeys. Generally, the audience provides monetary contributions at the conclusion of the performance. Although these performances are widespread across Asia, they also occur historically in Northern Africa and Europe as well (Janson 1952). Ohnuki-Tierney (1987, 1995) suggests that in Japan the monkey performance, and the macaque monkeys themselves, act as a mirror for humanity "playing a powerful role in their (humans') deliberations of who they are as humans vis-à-vis animals and as a people vis-à-vis other peoples" (1995: 297). As monkeys are the only animal addressed as "san," the adult human address, and referred to as "humans minus three pieces of hair," there is an aspect of social inclusion in Japanese culture that is distinct from the roles of other animals (Ohnuki-Tierney 1987, 1995). The integration of economic and cultural roles for macaques in Japan exists at the same time that increasing conflict over land and crops emerges as a predominant pattern of interaction between humans and "wild" macaques in Japan

(Sprague 2002). Again, as with pets and other economic uses, the direct impact of such inclusion on populations of macaques is similar to predation. However, the imposition of a broader cultural role for the macaques potentially impacts the macaque's position (or role) in the human *place*.

Temple or Urban Monkeys

All three of the above-mentioned situations can impact nonhuman primates, possibly affecting their behavior and physiologies by selectively removing individuals from breeding populations, exposing primates to new or different pathogens, providing novel food sources and contexts for interactions, or even by direct training.

However, none of the three have a broad, consistent impact on whole populations of primates. I suggest that the case of Asian temple macaques incorporates elements of the above three contexts but more directly reflects a domesticatory practice by humans. The relationship between humans and temple macaques is one that results in long-term and significant shaping of the lives and biology of the monkeys.

Throughout much of South and Southeast Asia at least three species of macaques appear to coexist well alongside human populations. Two of these species, the rhesus macaque *(M. mulatta)* and the long-tailed macaque *(M. fascicularis)* may be increasing their numbers in areas of direct habitat overlap with humans relative to forested or other areas considered "natural" habitat for these species. These increases are tied to human-directed changes to the local ecologies that provide favorable habitat or ecological features for the macaques, and to specific aspects of the macaques' "fit" within human cultural practice such that humans may also derive specific cultural benefits from the macaques (Fuentes, Southern, and Suaryana 2005).

Temple and urban monkeys exhibit extensive sympatry with human populations residing around specific temple structures in or near areas of human habitation. These can be in human cities, small local shrines and temples in villages and towns, or specific temple sites that double as sacred paces and tourist locations. This pattern is primarily in South and Southeast Asia and is strongly associated with cultures that practice Hinduism and/or Buddhism (Aggiramangee 1992; Fuentes, Southern, and Suaryana 2005; Zhao 2005). However, in the case of purely, or primarily, tourist locations, these patterns of interaction can also include nontemple tourist sites where monkeys are residents. Such tourist sites can be found in Gibraltar in Europe; many places in Kenya, South Africa, and Tanzania in Africa; and in Japan.

Although construction of Hindu and Buddhist temples in Asia reflect a time depth of no more than one to four millennia (depending on location) human and macaque overlap in these areas has a much greater temporal depth. Evidence for habitat overlap between macaques and modern *Homo sapiens sapiens* (and possible predation by humans) in Southeast Asia is clearly evident from remains at Niah Cave in Borneo over 25,000 years ago (Sponsel, Ruttanadakul, and Natadecha-Sponsel 2002). Given the prominent role of monkeys in Hindu and Buddhist myth (see below) and the high degree of security and nutritional assistance that humans provide to modern temple macaque populations, one can argue that there are human cultural benefits to this arrangement (Fuentes and Gamerl 2005; Fuentes, Southern, and Suaryana 2005; Zhao 2005). It is possible that this context and the relationship of sharing ecological space and cultural place acts as a domesticatory practice affecting both participants, humans and monkeys. This situation is exemplified by the pattern of interactions and coexistence between *M. fascicularis* and humans on the island of Bali, Indonesia

The macaques on Bali are found throughout the island, except in the extreme urban areas in the regency of Badung (such as the capital city, Denpasar). There appear to be minimally 63 sites where macaques reside on the island (Fuentes, Southern, and Suaryana 2005). Each of these sites has one to three groups of macaques that range either fully within the site or have the site as a part of their total home range. Each site has between fifteen to over 300 monkeys. Macaque densities at these sites range from one to over 20 individuals per square kilometer while human densities average over 500 individuals per square kilometer across the island. Most importantly, over 68 percent of these sites are associated with a temple or shrine. These religious complexes can be as small as a simple shrine consisting of a few stones and an altar to elaborate temple complexes that are heavily used by Balinese and in some cases foreigners (Fuentes, Southern, and Suaryana 2005).

An important component of Balinese Hinduism is the regular placement of offerings at shrines and temples. In most cases these offerings consist of at least 30 percent edible elements. This suggests that a large percentage of the macaque groups on Bali receive some substantial or integral component of their nutritional requirements from humans or human activity (see Fa and Southwick 1988; Wheatley 1999). Many of these temple macaque groups undergo some provisioning ranging from simply the consumption of the temple offerings to extensive daily provisioning with an array of fruits and vegetables by local temple staff. Provisioned foods can make up over half of the macaque diet at many temple sites in Bali (Fuentes, Southern, and Suaryana 2005). Presumably tied to this food availability temple site populations tend to have larger group sizes than nontemple site populations. Importantly, it appears that specific land-use patterns and wet-rice agriculture combined with the complex temple and irrigation systems of the Balinese (Lansing 1991) has resulted in an ecology that consists of a mosaic of riparian forest corridors and small forest islands throughout much of Bali (Fuentes, Southern, and Suaryana 2005). This type of landscape fits remarkably well with the macaques' patterns of using riparian habitats and small forest clusters for residence, foraging, and dispersal. This landscape has been formed over at least the last few millennia and the pattern of distribution of macaque populations across the island suggests that the macaques are exploiting it (Fuentes, Southern, and Suaryana 2005).

In this scenario the human niche construction, or creation of Balinese *place,* creates a habitat that is beneficial to the macaques via protection (when on temple grounds), nutrition (via provisioned foodstuffs), and ecology (connectivity across forested sites via riparian corridors). Simultaneously, the humans gain specific cultural and economic benefits from the presence of the macaques in and around their temple sites. The economic benefits include temple site entrance fees from tourists who come to see and interact with the monkeys and the additional funds that they spend in the local area on food, lodging, and the purchasing of goods. The cultural benefits may include a sense of "merit" (Zhao 2005) or a sense of connectivity with specific Hindu socioreligious narratives elicited via the daily interactions and/or provisioning of the macaques (Fuentes and Gamerl 2005; Fuentes, Southern, and Suaryana 2005; Wheatley 1999).

This intensive sympatry impacts the behavior and possibly other physiological facets of both primate species. There is also growing evidence of cross-species pathogen exchange and possibly coevolution to similar pathogen environments facilitated via mutual habitat use and frequent behavioral interactions (Engel et al. 2002; Fuentes and Gamerl 2005; Jones-Engel et al. 2005). It can be argued that this overlap, at least between macaques and humans, in parts of South and Southeast Asia at temple sites represents a domesticatory practice in which both humans and macaques are participants.

From "In-between" to "Inside": Nonhuman Primates' Role in Human *Place*

Human *place* involves cultural constructs and cultural behavior as well as a degree of manipulation of environmental and ecological parameters. Therefore, understanding the cultural context, or roles, for nonhuman primates helps us understand their inclusion in the human *place* and their potential participation in a human initiated domesticatory practice. Here we need to go beyond asking if the primates are impacted in morphological, behavioral, or physiological ways by human practice and ask how humans conceptualize nonhuman primates. These conceptualizations might affect the way humans incorporate the other primates into our active engagement with the environment *and* how humans might use primates as symbols within our own, or other's, cultural milieus (Lowe 2004).

Cultural views of nonhuman primates can range from conceptualizations of the primates as furry, little, near-people to demonic ambassadors to favored food stuffs to beguiling trickster deities and beyond (Corbey and Theunissen 1995; Fuentes and Wolfe 2002; Janson 1952). Of interest for this chapter is whether or not these conceptualizations result in the creation of a culture set wherein the nonhuman primates have a shared role in a human society's configurations of norms and patterns. If so, it is likely that they will be impacted in ways beyond simple coexistence.

Allopatric Contexts

In much of western Eurasia and North America primates play substantial roles in components of myth, folklore and popular imagery, especially as related to Judeo–Christian–Islamic myth and morality, political and social satire or defamation, and interpretations of what it means to be "wild" (Asquith 1995; Carter and Carter 1999; Corbey and Theunissen 1995; Janson 1952; Rothfels 2002).

Janson's (1952) *Apes and Ape Lore* and Corbey and Theunissens's edited volume *Ape, Man, Apeman: Changing Views since 1600* (1995) review integration of nonhuman primates into Euro-Western cultural mores and practice. Janson (1952) charts the ape's (read nonhuman primate's) change from *figura diaboli* to *imago hominis*, the last effort of creation before "man" sent to remind us "there, but for the Grace of God go I" (Janson 1952: 29). Janson painstakingly illustrates the patterns of representation of nonhuman primates in medieval and renaissance literature, art, and devotionals in Europe. From the role of *similitudo hominis*, doubling for our bodies, to the *vanitas*, folly and rampant sexuality (male and female) of the fallen primates, to the ultimate *Ars Simia Naturae*, art imitates life as the nonhuman primates are imitations, almost, but not quite, humans.

Emerging from these and other's works is a powerful picture of the allopatric conceptualization of nonhuman primates as our alter egos, somewhat less than human and somewhat more than animal. The morphological similarities and behavioral overlaps mixed with the rarity in actual exposure and the mythical creations of folklore imbued with Judeo–Christian–Islamic constructs create the space wherein the Eurocentric west envisions the nonhuman primates as the wild nature of humanity. This has resulted in their use in racist stereotyping during wars (see Asquith 1995), as elements in practices of disempowering minorities and ethnic "others," their use as caricatures of human vice, and their use as wily tricksters and infantile humans capable of being brought "up to our standards" (Carter and Carter 1999).

Linnaeus and other early taxonomists and scientists included the apes within our own genus. The recent enhancement in techniques of genetic investigation and the advent of primatology as a discipline has added a layer to the Eurocentric context for primates; viewing the primates as our close kin, and the apes as our brethren. From Lamarck's proposal to test the will to change via training an ape to Kafka's (1971) *report of a former ape to the members of a learned society,* authors have presented scenarios wherein the ape becomes the human (or almost so) because of their basal proximity to ourselves. This proximity and the implied potential for a "human soul" (however defined) has resulted in the proposal for the inclusion of a limited subset of primates, the so-called "great apes" (chimpanzees, gorillas, and orangutans) into human culture at the level of human rights (Cavalieri and Singer 1995).

Also stemming from this set of cultural perceptions of primates, humans from allopatric regions have begun using primates in defining the "wild" or "natural state of things." In some cases, such as Daniel Quinn's telepathic Gorilla in *Ishmael* (1995), the nonhuman primates can become "spokespeople" for the grand order of nature representing the natural path relative to humans' inordinate, negative, impact. Taken as a whole, the Euro-Western allopatric context can be seen as conferring a special status to the primates relative to other animals. This is a place near to humans, sometimes alongside humans. Primates are well integrated into Euro-Western cultural contexts not as mirrors but as basal, possibly degraded, or recently possibly more "pure," representations of our selves.

This allopatric perspective of theoretical familiarity and pseudokinship results in a particular set of interactions between tourists from allopatric zones and macaque monkeys at Asian temple sites. This set of interactions molds the behavior of the macaques. A substantial percentage of allopatric tourists exhibit a distinct pattern of behavior wherein they anthropomorphize the macaques extensively, seeing them as "little furry people," or other near-human entities (Fuentes 2006a). Such tourists often seek to offer food and achieve physical contact with the monkeys. They often become angered when the macaques do not conform to appropriate human social mores (such as sharing provisioned food or responding to verbal statements) and conflicts can result. These interactions can involve substantial physical contact and thus increased risks of pathogen transmission, potentially impacting both primate species' physiologies (Fuentes and Gamerl 2005). Therefore, in part because of cultural perceptions of primates, tourists engage in a set of behavioral interactions that results in the macaques learning and using specific behavioral repertoires (to acquire food or interact with humans) and increasing their risk of exposure to human-borne pathogens (Fuentes 2006a; Fuentes and Gamerl 2005; Jones-Engel et al. 2005).

Sympatric Contexts

There are diverse patterns across the sympatric distribution of human and nonhuman primate overlap. Rather than being generally classified as *imago hominis,* nonhuman primates across the zones of sympatry are frequently seen as distinct entities sharing commonalities with humans not as basal, or slightly degenerate forms, but, rather, as lateral relatives or avatars of humanity. In many cases there is a sense of fluidity in boundaries between the human and nonhuman primates (Burton 2002; Carter and Carter 1999; Pieterse 1995). Often there are also theistic aspects associated with monkey and ape such that the focal primates are components of polytheistic pantheons (although frequently as a semideity, or midlevel deity, rather than as primary deities—see Pieterse 1995; Fuentes 2006b). In a sense, the sympatric context can be envisioned as a reflection, or mirror (sensu Ohnuki-Tierney 1987), of humans broadly, not necessarily as imitators to humanity. There is also a theme of transformation of the monkey from a wild, or ambivalent to hierarchy character into a more learned and idealized humanesque character, such as typified by Sun WuKong's transformation in the Journey to the West (Burton 2002).

More recently, as the ownership and participation in primatological and conservation studies expands, we also see the that sympatric primates and their potential representation of "nature" and "natural" may be being co-opted by sympatric nation scientists in their postcolonial quest for definition and location on the international scientific stage (Lowe 2004). Celia Lowe (2004) suggests that these scientists may be using the presence and representation of certain sympatric primates (the Tongean macaque in her example) to simultaneously engage as elite (within their nation) and subaltern (within international science) in the utilization and contextualization of the primates at multiple levels (local, national, and international). In this emerging case we see a sphere of syncretism between allopatric and sympatric perspectives on primates.

In many sympatric regions, we see preferential treatment of some monkey species as they are considered ancestrally tied to the local cultures (e.g., Cormier 2002, 2003). Others because of their roles in prominent religious folklore and their association with religious structures or institutions are also given preferential treatment (Burton 2002; Fuentes, Southern, and Suaryana 2005; Wheatley 1999). In these cases, we see discrimination between either species of primates, or between primates and other mammals. These distinctions then, can act to establish differential positions for the nonhuman primates in the human's conceptualization, and construction, of their surroundings.

In regards to the temple macaques across Asia, two specific socio-religious narratives are especially relevant: the Ramayana and the Journey to the West. The Ramayana is a major Hindu epic that has roots at least 2,000 years old. One of the major roles in the Ramayana is that of Hanuman the monkey king (or general) who becomes a loyal and devout servant of Rama, the hero of the epic. Hanuman and his monkey legions facilitate the rescue of Sita, Rama's bride, from the demon lord of Lanka and the defeat of the demon armies. This epic and numerous affiliated folktales and myths are prominent throughout South and Southeast Asian societies, even those cultures that are not currently Hindu. Today, Hanuman, and his monkey associates and relatives, are considered to have beneficial value and are frequently tolerated in and around temples and other Hindu and Buddhist shrines (Wolfe 2002). Given the sincere devotion of Hanuman to Rama and the role that he played in assisting "man" combined with long-term exposure to monkeys by the human populations in South and Southeast Asia, we can see a cultural inclusion on both folkloric and practical levels.

The Chinese novel *Journey to the West* is generally presented in its most recent form (~16th century) but appears to have older roots. It reflects a confluence of Taoistic, Confucianistic, and Buddhistic elements, and may be influenced by aspects of the Ramayana (Burton 2002). The tale opens with the birth of Monkey (Sun WuKong) from a stone egg, who progresses to becoming the King of the Monkeys. Eventually, he creates chaos in heaven as he attempts to rise in the pantheon of deities. He eats the fruits of immortality specially grown for the Heavenly Queen Mother of the West, and upsets the Jade Emperor and other deities. He is then imprisoned beneath the Mountain of the Five Elements by the Buddha. Subsequently, he is released to accompany the monk XuanZang on his quest to obtain the holy Buddhist scriptures from India, along with three other pilgrims. On their journey they overcome calamities and confrontations in the form of supernatural phenomena and monsters before reaching their goal and returning to China with the texts. The story is loosely based on an actual journey undertaken by the monk XuanZang (600–664) to India, the home of Buddhism, to collect Mahayan Buddhist texts for translation into Chinese (Burton 2002). Although XuanZang is the ostensible focal point of the story, it is the exuberance, deviousness, jocularity, and innovation of the monkey that captures the lead role in the novel. In the end, Sun WuKong achieves a form of enlightenment earning the title "Buddha victorious in strife" (Burton 2002). This tale is extremely popular throughout much of Asia, especially China. The pervasiveness of this story impacts the conceptualization of primates in many of the human cultures where it is told. The role of monkey as assistant to humanity, but often engaging in chaotic, hierarchy challenging behavior at the same time, allows a fit for the interaction patterns between humans and nonhuman primates (esp. macaques) in much of Asia. The transition to maturity also allows for the cultural role of monkey as teaching tool, and therefore as valuable, to emerge.

At many temple sites in Asia, there exist explicit connections between the resident macaque species and the monkey participants in the Ramayana and the Journey to the West. Not that the macaque monkeys are considered "sacred" per se by such an association (see Wheatley 1999) but, rather, that they are tolerated and considered more culturally relevant than other animals because of the connection (Fuentes, Southern, and Suaryana 2005). The pervasive and ubiquitous presence of these myths and stories throughout much of Asia provides a cultural niche or role for the macaques that is not available to the same extent for other animals. This makes the temple macaques more likely to receive food, protection, and tolerance from humans. This relationship combined with the large-scale anthropogenic alterations in Asian habitats and the patterns of temple construction shapes the behavior, physiologies, and possibly the population genetics, of the temple macaques.

Are These Interconnections and Patterns "Domesticatory Practices"?

In this chapter, I have placed an emphasis on viewing human–nonhuman primate interconnections under the rubric of domesticatory practices. It is obvious from this brief overview of human–nonhuman primate interconnections and interactions that no single pattern of interaction can effectively characterize these relationships. However, some relationships, such as those between temple monkeys and humans in Asia, appear to be best characterized as involving domesticatory practices.

There are some fairly clear examples of relatively traditional domesticatory practices ongoing between humans and some monkeys. The coconut picking macaques and the primates used in monkey performances are good examples. This is a practical (functional or productive in an economic sense) process intentionally directed by humans to produce the desired outcome. We can also view the extensive sociocultural inclusion of nonhuman primates into the human's kinship system and frequent hunting as potentially having a molding impact on primate populations. However, the clearest example of a domesticatory practice comes from the cultural inclusion via myth and folklore combined with specific patterns of temple construction and subsequent alteration of the environment. This produces situations like that on Bali, or in other areas of Asia, wherein the human cultural and ecological patterns facilitate enhanced survivability and changes in behavioral patterns of the temple macaques. These situations result in contexts that impact the behavior and possibly the morphology of the primates concerned. They can also create pathogen environments that have an impact on human and nonhuman primate biology. The case of the temple macaques is an example of an ongoing domesticatory practice.

Expanding our gaze into the realm of cultural conceptualizations one can also argue that the allopatric Euro-Western incorporation of nonhuman primates from *figura diabolica* to *similitudo hominis* to *imago hominins* is a form of cultural inclusion that acts to control, co-opt, and integrate, the wild semihuman other. As Western scientific knowledge and methods increased, especially with the advent of intensive comparative anatomy, the clear distinction between human and animal becomes blurred (even before Linnaeus perceived this) and the figure of the allopatric primate becomes the focus for this quandary. The pattern of sociocultural inclusion in the Euro-Western tradition may act to resolve conflicts brought by notions of demarcation between humans and animals and the overwhelming evidence of commonalities between humans and other members of our order (Primates).

Where are the Relationships Today?

In both allopatric and sympatric contexts, nonhuman primates are commonly integrated into patterns and processes of how humans define who we are. Because of the similarities between humans and other primates, this is not surprising. I propose that there are four categories that help us envision the differing patterns and contexts that best characterize the relationships between humans and other primates today: integrated, engaged, penalized, and idealized or demonized.

Integrated—a traditional domesticatory practice wherein nonhuman primates are raised and selectively handled or trained to perform specific economic and/or nutrition enhancing tasks whose payoffs are collected by humans. This ranges from the training of coconut picking macaques to the large scale farming of pathogen free monkey colonies for biomedical investigation.

Engaged—wherein nonhuman primates are incorporated into human culture in a way that predisposes certain benevolent behavior by humans toward the primates. In addition to the cultural engagement, there is also an ecological engagement wherein the human modification of the environment, be it habitat restructuring, predation threat removal, or nutritional supplementation, provides a survivability benefit to the primates. This combination of cultural and ecological engagement as a domesticatory practice is best represented by the relationships between the temple macaques and humans in much of South and Southeast Asia.

Penalized—wherein deleterious interconnections form an antagonistic relationship between humans and nonhuman primates. This aspect has not been the focus of this chapter, but is self-evident for much of the discussion. Most of human niche construction is disruptive to other mammals, and hunting and collection by humans can have greater impact than that of most nonhuman predators. Habitat alteration and competition over space and food with humans penalizes most primates and results in population decreases and even local extinctions in many cases.

Idealized or Demonized—The cultural categorization of the nonhuman primates as degraded people or noble ambassadors of nature, and the "primatization" of human groups. These scenarios are played out in the discourse surrounding conservation ethics and ape (and all primate) rights, in the "othering" of human groups and, in the myriad of ways wherein primates are humans are symbolically connected or transposed in cultural contexts. In some sense, we can envision this category as the postmodern extension and replacement of the concept of the human noble (or ignoble) savage with the *imago hominins,* our primate cousins.

A dynamic view of the patterns and contexts assists in our attempts to model and understand the relationships between humans and other primates. Seeing these relationships within the context of human niche construction and possibly involving domesticatory practices allows for a greater understanding of some of the relationships between humans and the nonhuman primates. Assessing cultural conceptualizations in the light of sympatric and allopatric distributions and historical factors also enables greater understanding of the patterns of inclusion and representation of the nonhuman primates across human cultures. The interface between humans and nonhuman primates is a rich arena for discourse within anthropology, an arena in which our investigations are enhanced by embracing the complexity and dynamism inherent in these relationships.

References

Aggiramangee, N. 1992. Survey of semi-free ranging macaques of Thailand. *Natural History Bulletin of the Siam Society* 40: 103–166.

Alvard, M. J., J. G. Roninson, K. H. Redford, and H. Kaplan. 1997. The sustainability of subsistence hunting in the neotropics. *Conservation Biology* 11: 977–982.

Amman, K., J. Pearce, and J. Williams. 2000. *Bushmeat: Africa's conservation crisis.* London: World Society for the Protection of Animals.

Asquith, P. J. 1995. Of monkeys and men: Cultural views in Japan and the West. In *Ape, man, apeman: Changing views since 1600. Evaluative proceedings of the symposium ape, man, apeman: Changing views since 1600, Leiden, the Netherlands, 28 June–1 July, 1993,* edited by R. Corby and B. Theunissen, 309–326. Leiden, the Netherlands: Department of Prehistory, Leiden University.

Burton, F. D. 2002. Monkey king in China: Basis for conservation policy? In *Primates face to face: The conservation implications of human–nonhuman primate interconnections,* edited by A. Fuentes and L. D. Wolfe, 137–162. Cambridge: Cambridge University Press.

Carter, A., and C. Carter. 1999. Cultural representations of nonhuman primates. In *The nonhuman primates,* edited by P. Dolhinow and A. Fuentes, 270–276. Mountain View, CA: Mayfield Publishing.

Cavalieri, P., and P. Singer. 1995. The great ape project. In *Ape, man, apeman: Changing views since 1600. Evaluative proceedings of the symposium ape, man, apeman: Changing views since 1600, Leiden, the Netherlands, 28 June–1 July, 1993*, edited by R. Corbey and B. Theunissen, 367–376. Leiden, the Netherlands: Department of Prehistory, Leiden University.

Corbey, R., and B. Theunissen, eds. 1995. *Ape, man, apeman: Changing views since 1600. Evaluative proceedings of the symposium ape, man, apeman: Changing views since 1600, Leiden, the Netherlands, 28 June–1 July, 1993*. Leiden, the Netherlands: Department of Prehistory, Leiden University.

Cormier, L. A. 2002. Monkey as food, monkey as child: Guaja symbolic cannibalism. In *Primates face to face: The conservation implications of human–nonhuman primate interconnections,* edited by A. Fuentes and L. D. Wolfe, 63–84. Cambridge: Cambridge University Press.

———. 2003. *Kinship with monkeys: The Guaja foragers of eastern Amazonia.* New York: Columbia University Press.

Engel, G. A., L. Jones-Engel, K. G. Suaryana, I. G. A. Arta Putra, M. A. Schilliaci, A. Fuentes, and R. Henkel. 2002. Human exposures to Herpes B seropositive macaques in Bali, Indonesia. *Emerging Infectious Diseases* 8 (8): 789–795.

Fa, J. E., and C. H. Southwick, eds. 1988. *Ecology and behavior of food-enhanced groups.* New York: Alan R. Liss.

Fuentes, A. 2002. Monkeys, humans and politics in the Mentawai Islands: No simple solutions in a complex world. In *Primates face to face: The conservation implications of human–nonhuman primate interconnections,* edited by A. Fuentes and L. D. Wolfe, 187–207. Cambridge: Cambridge University Press.

———. 2004. It's not all sex and violence: Integrated anthropology and the role of cooperation and social complexity in human evolution. *American Anthropologist* 106 (4): 710–718.

———. 2006a. Human culture and monkey behavior: Assessing the contexts of potential pathogen transmission between macaques and humans. *American Journal of Primatology* 68: 880–896.

———. 2006b. The humanity of animals and the animality of humans: A view from biological anthropology inspired by J. M. Coetzee's *Elizabeth Costello. American Anthropologist* 108 (1): 124–132.

Fuentes, A., and S. Gamerl. 2005. Characterizing aggressive interactions between long-tailed macaques *(Macaca fascicularis)* and human tourists at Padangtegal Money Forest, Bali. *American Journal of Primatology* 66: 197–204.

Fuentes, A., M. Southern, and K. G. Suaryana. 2005. Monkey forests and human landscapes: Is extensive sympatry sustainable for *Homo sapiens* and *Macaca fascicularis* in Bali? In *Commensalism and conflict: The primate–human interface,* edited by J. Patterson, 168–195. Norman, OK: American Society of Primatology Publications.

Fuentes, A., and L. D. Wolfe, eds. 2002. *Primates face to face: The conservation implications of human–nonhuman primate interconnections.* Cambridge: Cambridge University Press.

Haraway, D. J. 1989. *Primate visions: Gender, race, and nature in the world of modern science.* London: Routledge.

Janson, H. W. 1952. *Apes and ape lore in the Middle Ages and the Renaissance,* vol. 20, *Studies of the Warburg Institute.* London: University of London.

Jones-Engel, L., G. A. Engel, M. A. Schillaci, R. Babo, and J. Froelich. 2001. Detection of antibodies to selected human pathogens among wild and pet macaques *(Macaca tonkeana)* in Sulawesi. *American Journal of Primatology* 54: 171–178.

Jones-Engel, L., G. A. Engel, M. A. Schillaci, A. L. T. Rompis, A. Putra, K. Suaryana, A. Fuentes, B. Beers, H. Hicks, R. White, and J. Allen. 2005. Primate to human retroviral transmission in Asia. *Emerging Infectious Diseases* 7 (2): 1028–1035.

Kafka, F. 1971. *The complete stories,* edited by N. N. Glatzer. New York: Shocken.

Lansing, S. J. 1991. *Priests and programmers: Technologies and power in the engineered landscape of Bali.* Princeton: Princeton University Press.

Leach, H. M. 2003. Human domestication reconsidered. *Current Anthropology* 44 (3): 349–368.

Lévi-Strauss, C. 1963. *Totemism.* Boston: Beacon Press.

Lizarralde, M. 2002. Ethnoecology of monkeys among the Bari of Venezuela: Perception, use and conservation. In *Primates face to face: The conservation implications of human–nonhuman primate interconnections,* edited by A. Fuentes and L. D. Wolfe, 85–100. Cambridge: Cambridge University Press.

Lowe, C. 2004. Making the monkey: How the Togean macaque went from "new form" to "endemic species" in Indonesians' conservation biology. *Cultural Anthropology* 19 (4): 491–516.

Mullin, M. H. 1999. Mirrors and windows: Sociocultural studies of human–animal relationships. *Annual Review of Anthropology* 28: 201–224.

Odling-Smee, J. F., K. N. Laland, M. W. Feldman. 2003. *Niche construction: The neglected process in evolution. Monographs in Population Biology,* 37. Princeton: Princeton University Press.

Ohnuki-Tierney, E. 1987. *The monkey as mirror: Symbolic transformations in Japanese history and ritual.* Princeton: Princeton University Press.

——. 1995. Representations of the monkey (Saru) in Japanese culture. In *Ape, man, apeman: Changing views since 1600. Evaluative proceedings of the symposium ape, man, apeman: Changing views since 1600, Leiden, the Netherlands, 28 June–1 July, 1993,* edited by R. Corbey and B. Theunissen, 297–308. Leiden, the Netherlands: Department of Prehistory, Leiden University.

Patterson, J., and J. Wallis, eds. 2005. *Commensalism and conflict: The primate–human interface.* Norman, OK: American Society of Primatology Publications.

Pieterse, J. N. 1995. Apes imagined: The political ecology of animal symbolism. In *Ape, man, apeman: Changing views since 1600. Evaluative proceedings of the symposium ape, man, apeman: Changing views since 1600, Leiden, the Netherlands, 28 June–1 July, 1993,* edited by R. Corbey and B. Theunissen, 341–350. Leiden, the Netherlands: Department of Prehistory, Leiden University.

Potts, R. 2004. Sociality and the concept of culture in human origins. In *The origins and nature of sociality,* edited by R. W. Sussman and A. Chapman, 249–269. New York: Walter de Gruyter.

Quinn, D. 1995. *Ishmael: An adventure of mind and spirit.* New York: Bantam Book.

Richardson, M. 1989. Place and Culture: Two disciplines, two concepts, two images of Christ, and a single goal. In *The power of place: Bringing together geographical and sociological imaginations,* edited by J. A. Agnew and J. S. Duncan, 140–156. Winchester, NY: Unwin Hyman.

Rothfels, N. 2002. Immersed with animals. In *Representing Animals,* edited by N. Rothfels, 199–224. Bloomington: Indiana University Press.

Shepard, G. H. 2002. Primates in Matsigenka subsistence and world view. In *Primates face to face: The conservation implications of human–nonhuman primate interconnections,* edited by A. Fuentes and L. D. Wolfe, 101–136. Cambridge: Cambridge University Press.

Sponsel, L. E., N. Ruttanadakul, and P. Natadecha-Sponsel. 2002. Monkey business? The conservation implications of macaque ethnoprimatology in Southern Thailand. In *Primates face to face: The conservation implications of human–nonhuman primate interconnections,* edited by A. Fuentes and L. D. Wolfe, 288–309. Cambridge: Cambridge University Press.

Sprague, D. S. 2002. Monkey in the backyard: Encroaching wildlife in rural communities in Japan. In *Primates face to face: The conservation implications of human–nonhuman primate interconnections,* edited by A. Fuentes and L. D. Wolfe, 254–272. Cambridge: Cambridge University Press.

Wheatley, B. P. 1999. *The sacred monkeys of Bali.* Prospect Heights, IL: Waveland Press.

Wolfe, L. D. 2002. Rhesus macaques: A comparative study of two sites, Jaipur, India, and Silver Springs, Florida. In *Primates face to face: The conservation implications of human–nonhuman primate interconnections,* edited by A. Fuentes and L. D. Wolfe, 310–330. Cambridge: Cambridge University Press.

Zhao, Q. K. 2005. Tibetan macaques, visitors, and local people at Mt. Emei: Problems and countermeasures. In *Commensalism and conflict: The human–primate interface,* edited by J. Paterson and J. Wallis, 376–399. Norman, OK: American Society of Primatology Publications.

"An Experiment on a Gigantic Scale": Darwin and the Domestication of Pigeons

Gillian Feeley-Harnik

From a remote period, in all parts of the world, man has subjected many animals and plants to domestication or culture. Man has no power of altering the absolute conditions of life; he cannot change the climate of any country; he adds no new element to the soil; but he can remove an animal or plant from one climate or soil to another, and give it food on which it did not subsist in its natural state. ... No doubt man selects varying individuals, sows their seeds, and again selects their varying offspring. But the initial variation on which man works, and without which he can do nothing, is caused by slight changes in the conditions of life, which must often have occurred under nature. Man, therefore, may be said to have been trying an experiment on a gigantic scale; and it is an experiment which nature during the long lapse of time has incessantly tried. Hence it follows that the principles of domestication are important to us.

—C. Darwin, *The Variation of Animals and Plants under Domestication*
(1868, 1): 2–3

Why should we care about domestication? Because the study of domestication will give us insights into the "conditions of life" that produce variation and change, the origin of species including ourselves—this is Darwin's answer. No matter how artificial, monstrous, or trivial, domesticated organisms may appear to be, they are "experiment[s] on a gigantic scale ... experiments which nature during the long lapse of time has incessantly tried." In Darwin's view, "varieties ... may justly be called incipient species." The key question is "How do varieties, ... incipient species, become converted into true and well-defined species" (1868: 5).

Darwin was at odds with most of his contemporaries in emphasizing the value of domesticates for studying the "conditions of life ... under nature." In the very passage cited above, Darwin noted that "M. Pouchet [1864] has recently insisted that variation under domestication throws no light on the natural modification of species" (1868: 2 n. 2). Cuvier (1822–23), Lyell (1830–33, 2: 36–48), Wallace (Darwin and Wallace 1859: 53–62), and Galton (1865) held the same view, albeit for different reasons. The International Code of Zoological Nomenclature still applies only to " 'natural' biological entities, and not to those created by human selection" (Johnston and Janiga 1995: 15). Yet most of the animals now used in laboratory experiments in the United States and Europe are "those created by human selection."[1] Domesticates have an anomalous status in current natural history as well.

Darwin acknowledged that "domestic" and "wild" were problematic categories. His unusual focus on human practices of domestication as a model and means of understanding variation and natural selection "under nature" offers us an opportunity to study the human–animal relations in domestication in England during the decades of the mid–19th century when historical divisions between the natural and social sciences were just beginning to emerge but were not yet entrenched. Darwin (1859: 20–42, 489–490) chose domestic pigeons as his "special group" in On the Origin of Species (hereafter, Origin) in part because the ubiquity and antiquity of pigeon breeding, then documented in the early Egyptian dynasties of some 5,000 years ago, came closer than any other form of domestication to bridging the gap between breeding in his own time and place and in Nature, the ever-present problem of scale that he described in Origin as "natural selection ... daily and hourly scrutinizing throughout the world" and "the long lapse of ages ... [the] long past geological ages" (1859: 84).

As I show, Darwin's decision to use domestication to understand transformations in nature also required him to cross over social boundaries between gentry such as himself and other classes of people, suggesting that classifications in contemporary natural history were part of larger patterns of social interaction. Darwin's notebooks, corresp-ondence, and references to the breeders in his published works, provide a kind of ethnography of domestication in his lifetime (1809–82) that I have pursued further through archival research (see Acknowledgments).

I have focused on the pigeon breeders of London who were among Darwin's main informants, especially the silk weavers of Spitalfields in the East End, urban handloom workers who exemplified the pigeon fancy for their contemporaries. In addition to producing silks of the highest quality in Britain, the Spitalfields weavers were known especially for the quality of their pigeons, for their love of singing birds, including many wild birds from the surrounding country, and for their bird market, the oldest and largest in London.

With growing recognition of the diversity and complexity of processes of domestication among and between humans and other creatures, several scholars across the four fields of anthropology argue that such historically variable dichotomies as "wild" and "domestic" should be seen in a larger context of human–animal relations, "with domestication as one human-animal relationship among many" (Russell 2002: 295; see Cassidy 2002: 124–173; Ellen and Fukui 1996; Ingold 1994; Mullin 1999), a process in which humans may also be transformed both socioculturally and physiologically (Flannery, Marcus, and Reynolds 1989; Leach 2003; Wilson 1988). Ingold (1996) and Terrell et al. argue further that domestication encompasses landscapes: "To understand domestication, what must be taken into account is not only the story of particular species, but also the whole range of species—the species pool—from which transformed species have been drawn, for it is not just singular species but landscapes that human beings have been domesticating since the dawn of human time" (2003: 328).

The purpose of this chapter is to answer questions arising out of this recent research. What is domestication in this case, and possibly "codomestication," and how exactly are these processes related to broader patterns of social, political, and environmental relations among and between humans and other living creatures that Darwin's problem suggests? How can we better understand the relationship between immediate generations of life and death and those distant in time and space? If "wild" and "domesticated" derive from local ideas and practices, changing over time as Darwin himself discovered, then how do they come to spread, or not, in everyday practice, and how do they come to form the basis of generalizations in the sciences and humanities, including anthropology?

I focus especially on the immediate generations involved in Darwin's still anomalous analogy in this chapter, rooted in still open questions about the significance of humans and "domestication or culture" in "the conditions of life." In the first two parts of the chapter, I examine how Darwin, among his contemporaries, came to focus on human processes of domestication as a way of studying what he saw as nonhuman processes "in Nature," and why he chose urban pigeon breeders and their pigeons as his conceptual, methodological, and social bridge across these divisions. The next three parts focus on how the pigeon breeders saw their circumstances, especially the house–workshops in which artisan–breeders and their birds made their common home, and the evident connections between their dwelling, weaving and breeding. I then turn to the domestication of landscapes in which Darwin and the breeders were joined, arguing that birds generally were key figures in representations of generational and biogeographical change in Great Britain during this time. I conclude by returning briefly to the question of how our conceptions of sociality among and between humans and other forms of life might be understood in broader terms.

"Pigeons if you Please"

Darwin's *Origin* begins with these words: "When on board H. M. S. 'Beagle', as naturalist, I was much struck with certain facts in the distribution of the inhabitants of South America, and in the geological relations of the present to the past inhabitants of that continent" (1859: 1). Darwin certainly used the facts he recorded in nearly five years (*Diary* 1831–36) traveling around the globe. Yet he was also quick to credit the insights he gained from his research on domesticated animals and plants. Darwin's long-standing interest in domestication is evident in his work from 1836 onward. Indeed, he drew the central metaphor of *Origin*—"natural selection"—not from the exotic places he had been, but from the popular culture of plant and animal breeders, farmers, and tradespersons in Great Britain and beyond, who informed him in the ways of "picking" and "roguing" they called "selection," or more generally "improvement." Although his inquiries ranged widely from horses to rabbits, dahlias, and cabbages, he concentrated most on pigeons. In the spring of 1855, almost 20 years after returning from his voyage on the H.M.S. *Beagle,* Darwin started raising pigeons in Down and traveling to London to meet with pigeon fanciers. Domestic pigeons became the "special group" for his analogy between "artificial" and "natural selection" in *Origin* (1859: 20–59).

Darwin's early interest in domestication is evident in the notebooks he kept from 1836 through the mid-1840s, covering subjects from horticulture, crossing and hybridization to various kinds of animals, plants, and birds (see, esp., "Questions and Experiments [1839–1844]" in *Notebooks*: 487–516; Secord 1985). Darwin's essays of 1842 and 1844, outlining his argument, both begin with an analogy between variation and selection in "domesticated" and "wild" organisms (*Essays*: 1–13, 45–84).

In spring 1855, Darwin followed the advice of William Yarrell, a newsagent in the parish of St. James in central London (as then conceived, later the West End), who suggested to Darwin that he study domestic pigeons (Darwin to W. D. Fox, 3/19/1855, *Corres.* 5: 288). Fancy pigeons were so various that some breeders (e.g., Dixon 1851) regarded them as separate species created by God. As Darwin wrote later in *Origin* (1859: 22–23), even "an ornithologist [if] told that they were wild birds ... would certainly ... rank [them] as well-defined species"; some kinds might even be seen as separate genera. However Yarrell, a self-trained naturalist, had recently completed his *History of British Birds* (1837–43: 260), the most comprehensive history yet written of birds in the British Isles, in which he argued that the Rock Dove *(Columba livia)* was their common ancestor; the author (Domestic 1841: 362) of the *Penny Magazine*'s survey of domestic pigeons concurred (see fig. 6.1). Through Yarrell, Darwin became acquainted with William B. Tegetmeier—a journalist, editor, and fellow naturalist—as well as other breeders in London. As Dixon's (1851) frontispiece to *The Dovecote and the Aviary* suggests, pigeon keeping had become an urban fancy by the early 1850s. Wright (1879: 3) later estimated that "nine-tenths of fancy pigeons are kept in towns" (see fig. 6.2).

By early May 1855, Darwin had bought a pair each of fantails and pouters (at 20s/pair) from John Baily and junior, father and son, well-known bird sellers in Mount Street, near Berkeley Square, not far from Yarrell's home and shop (Darwin to his son William, [25 April, 1855] and to his cousin W. D. Fox [23 May, 1855], *Corres.* 5: 322, 337). By 4 November, Darwin was encouraging Sir Charles and Lady Lyell to visit by promising that "I will show you my pigeons! which is the greatest treat, in my opinion, which can be offered to [a] human being" (*Corres.* 5: 492). Four days later, he was writing to J. D. Hooker that he had "pairs of nine very distinct varieties, & I love them to that extent that I cannot bear to kill & skeletonise them" (*Corres.* 5: 497).

By late fall 1855, he had been elected as a member of two London pigeon clubs, the Philo-Peristeron ("Lovers of the Dovecote" in Greek) and the Columbarium ("Dovecote" in Latin), and in early December, he had started writing to some 27 people in British colonies and elsewhere abroad asking them to send skins of pigeons and poultry long domesticated in those parts (*Corres.* 5: 509 n. 3, 510–511). After visiting Darwin in April 1856, and hearing how central the pigeons had

[*a*, Pouter Pigeon; *b*, Carrier Pigeon; *c*, Jacobin Pigeon; *d*, Ringdove; *e*, Rock Pigeon; *f*, Fan-tailed Pigeon; *g*, Nun Pigeon; *h*, Tumbler Pigeon. At top, several varieties.

Figure 6.1. A Rock Pigeon (e) on the roof of a barn, together with a few of some 25 kinds of fancy pigeons then recognized, and "house-doves" flying around the dovecote in the background (Domestic 1841: 361). The anonymous author notes that *Columba livia* ranges from "wild" to "domestic" to "emancipated descendants of our domestic breed," and that "all these varieties breed with each other, and with the wild rockdove; and without due care, all soon degenerate, as it is termed, and acquire the original form and colouring" (Domestic 1841: 362–363). This article was part of a series in *The Penny Magazine* 10 [1841] on animals, beginning with dogs, then camels, and ending with "the Weasel Tribe of the British Isles," with "The Domestic Pigeon" clustered among "Domestic Poultry," "Domestic Waterfowl," and "Hedgehogs."

Archangel.   *Tumbler.*   *Carrier.*   *Powter*

GROUP OF DOMESTIC PIGEONS.

Figure 6.2. "Group of Domestic Pigeons. Archangel. Tumbler. Carrier. Powter." Frontispiece to Dixon (1851), showing the pigeons on the rooftops of a smoky city with a weathercock on the church steeple in the background. Drawn on wood by Alexander Fussell, engraved by Jackson.

become to his questions about "natural selection," Lyell wrote to him on May 1st: "I wish you would publish some small fragment of your data *pigeons* if you please & so out with the theory & let it take date—& be cited—& be understood" (*Corres.* 6: 89).

Darwin kept working instead, writing to Tegetmeier on 24 June, 1856 that he now had 89 pigeons, and on 30 August that "I am crossing all my kinds to see whether crosses are fertile & for the fun of seeing what sort of creatures appear" (*Corres.* 6: 152, 210; see DAR 205 in Darwin Papers, Cambridge University Library, for Darwin's notes on embryology, hybridism, and sterility in pigeons). A year later, on 29 September, 1857, he wrote Tegetmeier saying, "With respect to Pigeons

I shall collect no more, for I think for my object of I have done enough" (*Corres.* 6: 459). In mid-June 1858, he began to write up his notes, finishing in mid-August and writing to Tegetmeier on 8 September: "At last, thank God, I have done with my Pigeons, & have just killed all the scores of cross-breds—I have about 22 pure-birds left, offering them to Tegetmeier who got them within the week" (*Corres.* 7: 154, 155 n. 3; appendix II). Ten years later he donated his collection of dead birds (60 domestic pigeons and six ducks) to the British Museum's Department of Natural History (*Corres.* 7: 217 n. 6).

Darwin (1859: 20–59) used domestic pigeon breeding as his point of departure for presenting his arguments about the origin of species through natural selection. The pigeons remained primary in his subsequent works on domestication, even compared to such "privileged species" (Thomas 1983: 100–120) as horses and dogs. Pigeons occupy over 100 pages in *Variation in Animals and Plants under Domestication* ([hereafter, *Variation*] 1868), compared to about 30 pages each on dogs and cats, 17 on horses and asses, roughly 15 each on pigs and cattle, eight on sheep, three on bees, and one on goats. Darwin's *Descent of Man, and Selection in Relation to Sex* ([hereafter, *Descent*] 1870–71), which he published three years later, is an analysis of "the descent or origin of man" as a process of domestication, drawing on comparisons of "savage and civilized" people with "wild" and "domesticated" animals already outlined in his *Journal of Researches* on the *Beagle* (1839, 1: 179), based in the common association, not limited to Darwin, of domestication with culture more broadly (see epigraph, p. 1).

"The High Value of Such Studies ... Very Commonly Neglected by Naturalists"

Domestication might seem the obvious way for Darwin to have re-envisioned the cosmos, given the historical importance of agriculture in the lives of people throughout the British Isles and the growing importance of pets in Darwin's lifetime. Thomas (1983) includes Darwin's argument that humans and other creatures are linked by descent as evidence for his thesis on the gradual emergence in England between roughly 1500 and 1800 of a "modern sensibility" concerning the relationship of people to the land, plants, and animals around them. "Without the long history of pet-keeping in England and without the knowledge accumulated through centuries of experience of domestic animals, it is hard to believe that the author of *Descent* could have made his case in quite the way he did" (Thomas 1983: 141–142).

Yet the idea of studying changes among "wild" creatures, or "natural" history, on the basis of domesticated animals and plants, identified with human cultivation, even civilization, was controversial in Darwin's time. Darwin himself noted that studies of domestication "have been very commonly neglected by naturalists" (1859: 4). Why was domestication so problematic that most of Darwin's contemporaries questioned its value as a model of organic life, and why did Darwin persist in his approach despite their views? Why did he focus on pigeons, rather than the sheep or cattle so critical to the British economy, the dogs or horses so dearly loved as pets, or for that matter, the cabbages or flowers of the gardens for which "the English" were becoming known through such works as Hippolyte Taine's (n.d./ca. 1871) *Notes sur l'Angleterre (1860–1870)*.

The two main arguments against domestication as a model of "conditions of life ... in nature" (see p. 1) were these. First, domesticates are "mere monstrosit[ies] propagated by art" that would not survive in the wild; Darwin's notebooks show that many breeders held this view (see, esp., *Notebooks* C: 133, where Darwin uses these words, and D: 107, and "Questions..." [487–516]). Second, domesticates demonstrate reversion to type, not diversification, the fixity of species rather than their mutability. Following Cuvier (1813), but without explicitly referring to Divine Creation, Lyell (1830–33, 2: 26, 36–43; see 32, 35) argued that "the best authenticated examples of the extent to which species can be made to vary, may be looked for in the history of domesticated animals and cultivated plants." Yet the fact that human beings have "reclaimed" just a few species from the "wild state" shows that "natural inclinations" to taming or domestication must have been "prospectively calculated and adjusted [in] the economy of Nature." Human beings cannot create domestic species by "force or stratagem." Furthermore domesticates' "limited variability" is "never transmissible by generation."

Galton's (1865) argument about the inherent dispositions of animals to tameness or wildness was essentially the same. In later revising his essay, Galton (1883/1865: 174–175) extended his analysis of domestication to include the education of human beings, and thus "civilization." In his own words,

[t]he finality of the process of domestication must be accepted as one of the most striking instances of the inflexibility of natural disposition, and of the limits thereby imposed upon the choice of careers for animals, and by analogy for those of men. ... [Animals] destined to perpetual wildness ... [a]s civilisation extends they are doomed to be gradually

destroyed off the face of the earth as useless consumers of cultivated produce. (Galton 1883/1865: 194)

Wallace (Darwin and Wallace 1859: 53, their two papers on trans-mutation that Lyell and Hooker read to the Linnean Society on 1 July, 1858) agreed with Cuvier and Lyell (and Galton's [1883/1865] later argument) that varieties produced by domestication have "strict limits, and can never again vary further from the original type, although they may revert to it" (1859: 53). Yet he argued that there were "essential differences" between wild and domesticated animals (1859: 59–60), and thus that "no inferences as to varieties in a state of nature can be deduced from the observation of those occurring among domestic animals" (1859: 61). In his view, "[t]he assumption, that *varieties* occurring in a state of nature are in all respects analogous to or even identical with those of domestic animals, and are governed by the same laws as regards their permanence or further variation[,] ... is altogether false." On the contrary, "there is a general principle in nature which will cause many *varieties* to survive the parent species, and to give rise to successive variations departing further and further from the original type [without 'any definite limits' p. 62], and which also produces, in domesticated animals, the tendency of varieties to return to the parent form" (1859: 54).

Darwin's views were different yet. He shared with Cuvier and Lyell (and Galton), the assumption Wallace found false, that "varieties occurring in nature are in all respects analogous to or even identical with those of domesticated animals, and are governed by the same laws as regards their permanence or further variations" (Darwin and Wallace 1859: 54), and shared with Wallace the idea that variations in nature were limitless. Yet he differed from all four in arguing that variations under domestication were also limitless, and that the analogy between domesticated and wild forms of life supported the mutability of varieties and the ongoing origination of species, even if his analogy based on the domestication of pigeons did not actually document the emergence of a new species from *C. livia*.[2]

Darwin's understanding of domestication and feralization was com-plex in at least three ways. The diary that he kept during his *Beagle* voyage (*Diary* 1831–36) shows that he (like Lyell 1830–33, 2: 45) had begun to think of tameness and wildness as variable, rather than fixed, states of being. As I argue elsewhere (Feeley-Harnik n.d.), I think the astonishingly "tame" birds and animals he saw on Chiloé in November 1834, and in the Galápagos Islands in September 1835, were crucial, not in the patterns of their anatomical variations, which he recognized only after returning (Sulloway 1982a, 1982b), but precisely because of their tameness (see *Diary*: 304, 305, 308, 310; *Notebooks* B: 4, 136; C: 50, 165, 189; D: 7, 71, 148; E: 103, 117, 174). His notebooks show that he began to think of the qualities of creatures as relative to their circumstances; what might be monstrous in one place or time might be adaptive in another (*Notebooks* B: 230; C: 4, 65, 66,85; D: 107). He was also aware that human ways of domesticating animals were diverse and changing. He commented at several points on "the quite recent progress of systematic agriculture and horticulture" in Britain, which dated to "barely a century" (*Essays* 1844: 51, adding at 84 n. 51 that "History of pigeons shows increase of peculiarities during last years").

Darwin clearly associated the latest form of domestication in the British Isles with British breeders' "extreme skill, the results of long prac-tice, in detecting the slightest difference in the forms of animals [with] some distinct object in view" and their "systematic" or "methodical selection" of preferred variants, while "roguing," or destroying, those that are not. Yet he also contrasted "blind capricious man" with an all-seeing "Being," in his Essay of 1844 (*Essays*: 50–51, 66–67), evocative of the then-proverbial eye of providence. In *Origin* (1859: 84, 469), the all-seeing "Being" became "Nature" or "natural selection," but no less watchful: "daily and hourly scrutinizing, throughout the world, every variation, even the slightest," "acting during long ages and rigidly scrutinising the whole constitution, structure, and habits of each creature, 'favouring the good and rejecting the bad."

As Darwin explained to his readers in *Origin*, he believed that "a careful study of domesticated animals and of cultivated plants," based on breeders' creation of variations through methodical selection, "im-perfect though it be," would provide the best "explanation [for] the coadaptations of organic beings to each other and to their physical conditions of life … the best and safest clue" (1859: 4). Secord (1981, 1985) argues that Darwin used his firsthand observations of breeders' fine distinctions of the pigeons' smallest variations, especially evident at shows, as a model of nature's scrutinizing eye (see fig. 6.3). At Thomas Huxley's (*Corres.* 7: 434–5) suggestion, Darwin (1868, 1: 134–77) included in *Variation* engravings of the birds that would give readers the kind of visual education that he got directly from the breeders. These images, "drawn with great care by Mr. Luke Wells from living birds selected by Mr. Tegetmeier," are to be compared with the wild rock pigeon *(C. livia)*, "the parent-form of all domesticated Pigeons," represented by

Figure 6.3. "Annual Pigeon Show of the Philo-Peristeron Society" at the Freemasons' Tavern, Great Queen Street, on 11 January, 1853, showing Almond Tumblers in the center and Carriers on the left, reported to rank first and second as fanciers' favorites. *The Illustrated London News* 22 [15 January 1853]: 37–38.

the dead bird that Dr. Edmondstone had sent him from the Shetland Islands (see figs. 6.4 and 6.5).

For biologist Mary Bartley (1992), Darwin's pigeon breeding was not simply a metaphor or analogy for natural selection. More than any other of his experiments, pigeon breeding gave him the data on inheritance and growth for his theory of generation called "pangenesis," outlined at the end of *Variation*. Indeed, in *Variation* (1868, 1: 3), as indicated in the epigraph to this paper, Darwin seemed to shift his perspective from seeing the relation as an "analogy," as he did in *Origin*, to seeing it as an "experiment": "Man … may be said to have been trying an experiment on a gigantic scale … an experiment which nature during the long lapse of time has incessantly tried."

In *Origin* (1859: 31), Darwin acknowledged that "Breeders habitually speak of an animal's organisation as something quite plastic, which they can model almost as they please. … Youatt [a well-known English breeder] … speaks of the principle of selection as 'the magician's wand, by means of which he may summon into life whatever form and mould he pleases.' " In *Variation* (1868, 1: 234), Darwin described the pigeons

Fig. 19.—English Carrier.

Figure 6.4. "English Carrier" (Darwin 1868, 1: 135, 140), drawn on wood by Luke Wells, engraved by Butterworth and Heath.

as "mere stones or bricks ... without the builder's art." Yet to make his argument that variations derive randomly from unknown causes, while human selection may include unanticipated results, he downplayed the very artisanry he admired, describing the breeders' work in *Origin* as "a kind of Selection, which may be called Unconscious" (1859: 189), lest he imply that nature is governed by a grand design (1859: 34; 1868, 1: 226).

Thus, we are left with a curious paradox. Darwin (1859: 4) stated that the main value of domestication as a model of natural selection, and a means of understanding it, was that variation under domestication helped to elucidate "the coadaptations of organic beings to each other and to their physical conditions of life." He (Darwin 1859: 485–486) often drew on earlier views of the "economy of Nature" (e.g., Lyell 1830–33, 2: 43 cited above), even imagining the "production" of organic beings as if they were the "mechanical invention[s] ... of numerous workmen." Yet he never considered the coadaptations of humans with other creatures in these broader terms, either their working relations or their social relations more broadly. So I propose that we take Darwin's

Diagram of Carrier.   61

POINTS OF THE CARRIER.

Figure 6.5. "Points of the Carrier," showing key measures for evaluating quality (Fulton 1874–76: 61).

work on the breeders' practices of domestication as an entry into what he left out. Let us examine what his German contemporary Ernst Haeckel (1866)—inspired by Darwin's setting of mutual relations in the economy of nature—called their "Oecology [from] *oikos*, house/hold, living relations" (in Stauffer 1957: 140); and what his U.S. contemporary Lewis Henry Morgan (1881)—focusing on the kinship and affinity of social beings, not their functional relations as building materials—would have called their "houses and house-life." I argue that this approach, explicit in the work of Darwin's contemporaries outside of Great Britain, but supported by the similarities that Darwin and the breeders alike observed between breeders and birds, will help to elucidate processes of codomestication in the context of their common homes set in the urban ecology of London at this time.

"Little Men"

The London pigeon breeders whom Darwin got to know best were in the trades—newsagents, journalists, illustrators, tailors, and brewers, among others. Their businesses were in central London and in Southwark; their clubs in pubs in the areas of Covent Garden and Lincoln's Inn Fields, Berkeley Square, the Borough (Fulton [1874–76]: 384–86).[3] Darwin respected them for their knowledge, yet described them as "little men" in a letter to his son William on 19 November, 1855 (*Corres.* 5: 509): "I am going up to London this evening & I shall start quite late, for I want to attend a meeting of the Columbarian Society, which meets at 7 oclock near London Bridge. I think I shall belong to this Soc.y. where, I fancy, I shall met a strange set of odd men. Mr. Brent was a very queer little fish. ... after dinner he handed me a clay pipe, saying 'here is your pipe' as if it was a matter of course that I shd. smoke. Another odd little man (N.B. all Pigeons Fanciers are little men, I begin to think) & he showed me a wretched little Polish Hen, which he said he would not sell for £50 & hoped to make £200 by her, as she had a black top-knot."

Darwin's correspondence and the London Post Office directories of the time give us some glimpses of these men. They also wrote about their interests, or were written about by their kin or their friends. We know them mainly through their birds. Their writings about pigeons and pigeon breeding, and their drawings and paintings, often portraits, give some indication of what Darwin left out of his account, especially concerning the association of pigeon breeding, domestic relations, and their own homes and yards.

Pigeons were celebrated first of all for their loyalty and devotion. Their spiritual significance in biblical texts was increasingly discerned in the ancient texts and contemporary practices of Muslims and Hindus. The breeders described them as "monogamous," expressing their love for each other in "the Colombine kiss" and forming lifelong unions. They were extraordinarily fertile, creating pairs of eggs almost every month of the year. The male and female birds together brooded over the eggs, then nursed their young with "Pigeon's Milk," which, in being regurgitated from their crops, seemed to come from their breasts. The pigeons' loyalties to their mates allowed a breeder to keep pairs of many varieties in the same loft without worrying that either would step into another's nest. Yet they were also communal birds, preferring to live their paired lives in flocks; their social bonds were thought to explain their famous "homing" ability.

Second, fanciers seem to have treated their pigeons, at least their prize birds, as children whose presence could evoke their own childhoods, and perhaps vice versa: during his pigeon-keeping years, Darwin took to calling his son Leonard by the nickname "Pouter" (*Corres.* 7: 158, 159 n. 5). Breeders' favorite pigeons were typically the kinds they played with as children, remembered together with the parent (usually the father or grandfather) or family friend who gave them their first birds, and almost always with a close childhood friend. Although Darwin abstracted the fanciers' designs and social relations from his analysis of the historical origins of the various breeds of pigeons, the fanciers connected them, representing the birds' genealogical histories as human genealogies.

Third, they were linked to ideals of craftsmanship. Although Darwin downplayed the breeders' artistry, this was the breeders' main interest. The art in pigeon breeding lay in the beauty of the birds—their forms, feathers, and shimmering iridescent colors—but also in the delicacy of the relations through which they agreed to make their home with the breeder, be fruitful and multiply. In the words of Reverend Lucas, writing his memoirs as a pigeon fancier in 1886, "Pigeon-Fancying is the art of propagating life" (1886: 14). These ideals seem to have differed in different trades. Racing pigeons became the favorite of miners in the north of England. In the case of fancy pigeons in England, the trade was weaving. The enormous importance of these craft ideals may be better appreciated if we go even deeper into the heartland of the pigeon fancy, which—as everyone knew then—lay not with the middling classes of newsagents, journalists, and better sorts of tailors, but with people formerly called "artisans," but then increasingly "mechanics" or "operatives," and not in central London, but in the East End. As the Reverend Lucas (1886: 34) also explained: "Considering the pedigree in England, Spitalfields is the cradle of the fancy."

"Spital-field Weavers & All Sorts of Odd Specimens of the Human Species, Who Fancy Pigeons"

On 29 September 1856, Darwin wrote to a U.S. colleague: James Dwight Dana, a geologist and zoologist at Yale University: "In the case of Pigeons, we have (& in no other case) we have much *old* literature & the changes in the varieties can be traced. I have now a grand collection of living & dead Pigeons; & I am hand & glove with all sorts of Fanciers, Spital-field weavers & all sorts of odd specimens of the Human species, who fancy Pigeons" (*Corres.* 6: 235–236). Darwin would have learned about the fanciers of Spitalfields from the men he had met through Yarrell. The origins of the fancy among artisans and mechanics is clear from their accounts. For example, the same Mr. Brent, whom Darwin found to be "a very queer little fish" in 1855, had written a couple of years earlier: "Time was, and not many years since, when a Pigeon Fancier was associated in all men's minds with Costermongers, Pugilists, Rat-catchers, and Dog-stealers, and for no other reason that we can discern than that the majority of Pigeon Fanciers were artisans—men who lived in the courts, alleys, and other by-places of the metropolis" (1853: 35).

Except the Bailys' shop near Berkeley Square, most of the London bird markets were located in the poorest neighborhoods, like St. Giles in the Fields and Seven Dials west of Covent Garden Market, Whitechapel in the East End, and St. Olave's in Southwark. Of these, the best known was the Club Row market in Spitalfields north of Whitechapel. This is evident from many sources, for example, Mayhew's (1849) letters on "Labour and the Poor," first published in the *Morning Chronicle*, beginning with the Spitalfields weavers, then continuing with letters on street sellers and costermongers, the dog and bird sellers foremost among them. The Club Row bird market, extending into Sclater Street, was located next to the Spitalfields Market behind Shoreditch station, the terminus of the Eastern Counties Railway built there in 1840. This market remained the most important bird market in London up to 1983, when the R.S.P.C.A. succeeded in banning the street sale of live animals, after decades of protest (see fig. 6.6).

Who were the Spitalfields weavers? And why were they such "odd specimens of the Human species [that they would] fancy Pigeons" (to use Darwin's words)? Besides the accounts of contemporary writers like Mayhew, medical men like Dr. Hector Gavin, missionaries in the London City Mission Society, and others, the most detailed evidence comes from Parliamentary Papers (P. P.), especially the reports of the Select Committee on Hand Loom Weavers written in 1838–41, when Darwin was writing his notebooks on "transmutation" and "metaphysical enquiries" (see P. P.).

Spitalfields was a place just outside the old east wall of the City, long identified with the manufacture of silk, by handloom weavers in families, living in terraces with tiny yards, doing work so exacting that it required what was then called a "weaver's eye." During the years that Darwin was surveying on the *Beagle* (*Diary* 1831–36), writing his notebooks and essays (1836–44), and then *Origin* (1859), they had become the emblem of the most intense debates in political economy: over the persistence of handloom weaving in the face of the rapid industrialization of the textile trades, the value of international "free" trade in promoting domestic wealth; whether the Spitalfields weavers could complete with other districts without the protections of the

Figure 6.6. "The Bird Mart," Shoreditch, drawn by E. Buckman and engraved by W. Thomas for *The Graphic* 1 [25 December 1869]: 76. The engraving shows weavers' "lights" (long windows) and pigeon cotes at the tops of some houses and, among the bird sellers, a sailor from the London, St. Katharine's, or East London Docks south of Shoreditch, where silk workers sought jobs in hard times.

Spitalfields Acts, rescinded in 1826; in short, what one of the reporters on the conditions of the handloom weavers called "the great mystery of human progress" (P. P.–Fletcher 1840: 187). Drawing on the reports of these contemporary observers of the "Spitalfields weavers" in the process of their transmutation in the 1830s–1850s, that eventually resulted in what became known as the "black decades" of the 1860s–1890s (Warner ca. 1921: 542), I focus on three key issues: (1) land and house–yard–gardens; (2) the nature of the work requiring a "weaver's eye"; and (3) its domestic locus and mode of descent.

"Plain, Figured, and Fancy Articles"

Spitalfields, the fields east of the old City of London that later became the parish of Christchurch, Spitalfields, was as J. Mitchell reported to the Select Commission: "the oldest seat in England of the silk manufacture"

(P. P.–Mitchell 1840: 52). Legal and other documents from the 1330s–1600s testify to the presence of "the Trade Art or Mystery" of Silkworking in London since the 1330s (Warner ca. 1921: 626–631). Owing to their delicacy, silk fibers were liable to breakage. Thus, handloom weaving persisted much longer in the manufacture of silk cloths, compared to cottons, wools, and linens, and in all three branches of the trade: "plain," "figured," and "fancy"—"fancy" being the technical term for the "highest class of figured goods," including velvets, cloths of gold and the like (P. P. 1831–32: 8395).

"It being a Fancy Article," to cite the testimony of a silk manufacturer to the House of Lords in 1823, "the Capital, like a Nursery Ground, rears the Articles and propagates the Manufacture to all Parts of the Kingdom" (P. P. 1823: 177). As a "Fancy Article," the silk manufacture was at once highly lucrative, yet highly seasonal and, despite intermittent protective tariffs, ceasing in 1826, it was utterly unpredictable, subject to "caprices of fashion [that] baffle all calculation" (P. P.–Mitchell 1840: 215). Thus, weaving in Spitalfields was associated with the development of market gardening, seasonal labor at the docks on the north side of the Thames, and street selling (P. P.–Kay 1837: 192).

The transmutation of the silk manufacture in relation to the changing ecology of London is most evident in land and housing. Silk weaving in the reigns of Queen Anne (1702–14) and King George I (1714–27) was still based in house–workroom–shop–yards in the parish of Christchurch, Spitalfields, connected to patches of garden land and garden houses out in the hamlet of Bethnal Green or further east (see fig. 6.7). As the silk manufacture spread into Shoreditch, Bethnal Green, and neighboring parishes, the townhouses gave way to densely packed rows of houses with tiny yards and rare garden patches that were even more critical to supplementing income as some of the more easily mechanized aspects of the silk manufacture moved out of London (P. P.–Hickson 1840: 652, 655), driving prices down in Spitalfields. Thus, the pigeons and other domestic and wild birds, plants, and produce sold in street markets must also be seen as the outcome of a kind of radical intensification of the historical process of enclosure, although not officially subject to the Enclosure Acts, and in an urban setting (see fig. 6.8).

The weavers' birds and plants alike were clearly tied to artisanal ideals. They were an aesthetic expression of ideals of craftsmanship associated with silk weaving, especially skills in the design and execution of the forms, textures, and patterns of "fancy" goods, like velvets, damasks, and patterned silks, in which the "weaver's eye" was crucial to the quality of the work. So when Huxley (1862: 101), in his "Lectures for

Figure 6.7. Lamb Gardens, Bethnal Green, circa 1870, showing garden houses turned into weaver's house–workshops. Photograph by the Improved Industrial Dwellings Company, which had just bought the land and was about to raze the houses and construct the Waterlow Buildings, named for Sydney Waterlow who founded the commercial philanthropy in 1863. Reproduced with the permission of the Tower Hamlets Local History Library and Archives (331.1).

Working Men" in London, introduced Darwin's theory using Darwin's example of pigeon breeding, yet emphasizing his ignorance of the "great art and mystery of the pigeon fancy," he may have used the guild language of their craft to convey his seriousness, but he was also expressing the connection between their craft and their fancy. Thus, the scrutinizing eye of the fanciers, so apparent in the handling of their birds, has its corollary, perhaps its prototype, or more likely its interactive counterpart, in the scrutinizing eye required in the most skilled work, the Fancy Articles, of their craft.

The extraordinary craft expressed in the "weaver's eye" can best be appreciated from a closer examination of the cloth itself, for example, the "coronation velvet" that George Dorée—a velvet weaver and warp spreader—wove for the coronation of Edward VII, held on 9 August,

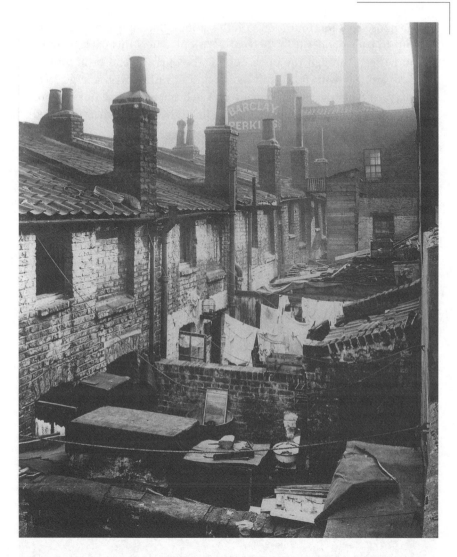

Figure 6.8. Backyards of houses on Pereira Street, looking south from 9 Holmes Ave, circa 1900, razed about 20 years later, after WWI. Note bird cages in two of the backyards (on wall of privy, at kitchen door), and pigeon dormer in the yard at the end of the row. Reproduced with the permission of the Tower Hamlets Local History Library and Archives (331.1).

1902. Frank Warner, a silk manufacturer, interviewed George Dorée in his home in Alma Road in 1914, two years before Dorée died. Here is his account of Dorée's house and yard (destroyed in 1959 to build the Cranbrook Estates in Globe Town):

The house contains four rooms on the ground floor, and a passage from front to back divides it in the centre. As one enters this passage, there can be seen through the open door at the opposite end of it, a small back-yard, gay with flowers in bloom and furnished with a large, neat aviary, in which a few specimens of a delicate prize breed of pigeons coo and strut in the summer sunshine in all the pride of their pencilled iridescent plumage. (ca. 1921: 99; see fig. 6.9)

The entire upper floor, reached through a trapdoor, was occupied by Dorée's warp-spreading machinery and looms used by other family members (see fig. 6.10). Therefore, he wove the coronation velvet in his parlor. According to what George Dorée told William Manchée,

Figure 6.9. Back of George Dorée's former residence at 42 Alma St. (looking south from Cranbrook Street), 22 February, 1933. Photograph by William L. Restall J. P., with notes on the location of the pigeon dormer on the left and the replacement of the old weavers' lights with smaller windows at the fronts and backs of the houses. Reproduced with the permission of the Tower Hamlets Local History Library and Archives (331.1)

Figure 6.10. George Dorée next to warp-spreading machine, with bird cage at back window, 1909. Two large looms were located along the front windows. Reproduced with the permission of the Tower Hamlets Local History Library and Archives (680.2).

who interviewed him for the Huguenot Society some months before his death in 1916:

> [King Edward VII's] Coronation velvet was thirty yards long and twenty inches wide [Queen Alexandra's velvet was twelve yards long]. To use the technical expressions of the trade, it was an 1850 thread, 60 wires, 180 shoots, and treble pole with a two-thread ground. It took six yards

of the top cane [warp] to make one yard of the pile of the velvet. [This means] that Mr. Dorée ... had to weave in the brass reed sixty times to every inch; nevertheless the cut of the pile is so true that the grain of the pile cannot be seen. ... To each square inch 33,120 threads stand on end, to each yard 23,846,400 or 763,084,800 threads in the entire piece, and withal perfect in match and size, so strong that a man could stand on it without pressing it down. (1918: 420–421)

This skilled work was transmitted through families; the "descent" of families was inextricable from the descent of their craft in their part-icular working conditions. Men carried out their transactions with other men, including their trade in birds, in public houses, beer shops, and markets. The transmission of their craft, like the breeding of their birds, was clearly familial, based in their home–workshop–yards, involving spouses and children, and passed on generationally, a point that comes up repeatedly in the testimonies of weavers and manufacturers. According to the testimony of John Duce, a Spitalfields weaver cited in Mitchell's report:

> Too great attachment to the occupation is the bane of the trade. ... There is a sort of independence about the work. ... The weaver also enjoys the society of his family. He feels strongly the domestic attachments, even beyond what cool reason would approve. Hence he will not find any other employment for his children, and easily believes that he cannot. From all these causes it is, that many are drawn into the trade, and after-wards bring up their children to it, and cling to the trade under every disadvantage. (P. P.–Mitchell 1840: 77)

The work could be inherited through objects like looms and their mechanisms, pattern books, and other tools of the trade. But it was also understood by participants and observers alike to be transmitted through the body in at least two ways. First, in the transmission of skills beginning in childhood, expressed as "bring[ing] up their children to it" (P. P.–Mitchell 1840: 376). Second, the descent of a trade was understood in bodily terms in the parliamentary papers. The silk weavers of Spitalfields were widely known as "little men," because of the very practice of their craft over generations. So Mitchell reported "That the weavers of Spitalfields are of a small stature has long been matter of public notoriety. ... Many of the witnesses examined have stated, that they were sons and grandsons of weavers, and that their wives were the daughters and grand-daughters of weavers. Such men form a caste, and are of small size" (P. P.–Mitchell 1840: 70–80). Mitchell quoted the observation of a weaver, Mr. Redfarn, to which others agreed: "They are decayed in their bodies; the whole race of them is rapidly descending to the size of Lilliputians." William Bresson, a velvet weaver and loom broker who testified before Hickson (P. P.–Hickson 1840: 717), also emphasized this point.

Finally, I argue that these ideas and practices of the hereditary craft of silk weaving—so deeply embedded in the constitutions of weavers that they passed into the bodies of their children—were expressed in their birds. The pigeons also involved a kind of "bringing up" or "shaping" of children, using the same skills and ideals as in the silk-weavers' craft, in which the shaping of children who were to be the next generation of workers was essential to the ongoing life of family and trade alike. As among the middling classes of breeders, the pigeon fancy in Spital-fields seems to have been a lifelong pursuit, passed on from grand-fathers to fathers to sons. These relations were especially important in the fancy branch, given the importance of the "living draw-boy," by comparison to the mechanical drawboy, then being introduced in the mechanization of fancy work. As Warner put it—he having moved the silk manufacture inherited from his father from Spitalfields to Braintree (Essex) in 1895—"in a few years, at most, the silk-weaving industry in London will become extinct for lack of weavers, [leaving] nothing quite like the methods and traditional arrangements of Spitalfields [to be] found elsewhere" (ca. 1921: 100). Indeed that happened. Their pouter pigeons were bought up by Scotch fanciers in Glasgow, Dundee, Edinburgh, and Paisley, where the trade in fancy scarves, which had become their mainstay, was relocated (Fulton 1874–76: 96).

"The Discrimination of Birds"

[W]e have often noticed, as others must have also done, what may be called the *discrimination* of birds, in relation both to persons and to places. We allude to what we should call their accommodating rather than their natural instincts, how, for example, after a season or two of observation or experience, they will congregate around a spot where no rude hands disturb their mossy dwellings, nor climbing urchin shows his visage grim among the umbrageous boughs.

—A. Newton, "Ornithology," *Encyclopaedia Britannica* 16 (1858): 793

The weavers' birds exemplify ideals specific to the silk industry, but also more broadly shared among artisans in textile manufactures in

Britain, because of shared ideals about craftsmanship especially in fancy work; because of shared experiences of the political–ecological circumstances involved in urbanization; and possibly because of shared experiences of the fragility of their familial lives during these years of rapid industrialization in which the hereditary transmission of the trade from parents to children was ever harder to achieve; and the possibility of keeping their children under one roof, or even within the British Isles, was difficult. These changes were expressed in what could be called an avian imagery of biogeographical change, evident in the changing places of birds in rural and urban settings and in contemporary slang, especially about migration and resettlement.

Domestic pigeons had been a mainstay of rural agriculture, a legal monopoly of British nobility and gentry, beginning in the mid 1600s, manifest in large, free-standing dovecotes containing 500 to 1,000 nesting holes, built or rebuilt in the years from 1640 to 1750, some still standing (Cooke 1920). Manorial dovecotes were one element in the diversification of "basic grain-livestock farming" to include "fruit vegetables, hops, and industrial crops," as well as small stock and game (Thirsk 1992: 51–52). Rising grain prices after 1760 encouraged the adoption of a new agricultural system for integrating crop and large livestock production based on turnips and swedes (rutabagas), clover, cattle, and sheep (Holloway 1996: 27). Because livestock could now be fed cultivated crops, especially turnips, over the winter, pigeons were no longer critical for winter meat. As the monopoly on pigeons was no longer rigorously enforced, "house-pigeons" were taken up by small farmers and laborers for their own subsistence and local sale.

After a lull between 1640 and 1750, landowners began a new series of enclosures lasting into the 1870s, creating larger estates, expelling unwanted laborers, and destroying their cottages (Thirsk 1992: 51–52). The enclosures in 1750–1870 affected mainly farms in the English Midlands and in Ireland, where English and Scottish settlers owned 95 percent of the land by the late 1700s. By the 1850s, "there was very little arable left in open fields although upland commons remained." By the 1870s, 70 percent of farmland in England was owned by 13,000 landlords (Holloway 1996: 27, 30–33). In ways I cannot discuss here, "British Birds," as Yarrell (1837–43) called them, became salient markers of ecological transformations associated with these changes in land use and ownership.

The number of people living in cities in England and Wales increased from 34 percent to 54 percent of the population between 1801 and 1851, continuing to rise steeply thereafter (Thrift and Williams 1987: 28).

Avian idioms of migration and urban resettlement, are strikingly evident in slang applied especially to the poor who migrated: the "rookeries" where they congregated, the "bird cages" or slop houses built there (as the building industry became the second largest industry in Britain after textile manufacture); the "call birds," or model-houses built to lure them in; and the "birds of prey" that hovered around them (to use Dickens's words in *Our Mutual Friend* [1864–65], esp., bk. 1, ch. 1: 3–4, 13–14 passim; bk. 1, ch. 17: 231; bk. 2 "Birds of a Feather," ch. 12 passim). Precisely because of their well-known capacities for "discrimination," the birds served here—as anthropologists have documented elsewhere (e.g., Feld 1987; Friedrich 1997; Reina and Kensinger 1991)—to span minutiae and immensities, far and near, that might otherwise have seemed incommensurate.

House- or dovecote pigeons, considered the closest descendants of the Rock Pigeon in the British Isles, were among the small stock that rural migrants brought into cities for food. By contrast, fancy pigeons were thought to have originated in the Near and Far East—Turkey, Palestine, Egypt, Arabia, India, Persia, and China—from which (like raw silk) they were now being imported. "From Egypt direct to Birmingham," as Fulton (1874–76: 337) said of the pigeon trade in Britain's second largest manufacturing center after London. For Darwin, the pigeons were both an image of understanding *and* a means of understanding vast changes experienced abroad but also at home. Darwin seems to have come upon the pigeons as a way of understanding life processes precisely because of the ancient and global scope of their close relations with humans. As he said in *Origin*: "[P]igeons have been watched and tended with the utmost care, and loved by many people. They have been domesticated for thousands of years in several quarters of the world" (1859: 27).[4] In Spitalfields, and I think more broadly in the British Isles, birds were intimately linked to the changing political–ecology of rural and urban life, including processes of manufacturing in the textile trades, in which design—that is, fashions in consumption—increasingly dictated the seasonal rhythms of weavers' work. For the silk weavers of Spitalfields, pigeon breeding was a kind of handloom weaving in the very materials of life.

Darwin's (1859: 130, 489) argument about natural selection in *Origin* is increasingly biogeographical and ecological, moving from a "great Tree of Life" to an "entangled bank, clothed," as he says, with diverse kinds of plants, birds, insects, and worms, so different and so complexly interdependent. Yet Darwin's argument about artificial selection is strikingly abstract: Artificial selection is to be found in the eyes, but especially the hands of breeders, taken from context. Indeed, the larger context of interests in "domestication" in the last century in Great Britain, intensified with the development of the Wardian case for the global transport of plants in the late 1830s to early 1840s (exactly those years of Darwin's notebooks and the parliamentary papers), had been to take organisms from one country and "domesticate" or "naturalize" them in another. These processes by which the birds moved from tangled banks to hands, and from fields to rooftops, suggest that Darwin's vision of the "the hidden bond of descent," conceived in genealogical terms, may indeed link all forms of life more tightly, yet more unequally. Historical processes of domestication through abstraction from one place and relocation in another, dispossession, migration, and resettlement, enabled some people to gain more control over the reproduction of themselves, their families, and their livings, while others clearly lost such control. The ostensible delinking of livestock and land in the contemporary industrial livestock sector, masking the global geopolitical restructuring of large firms controlling food production from reproduction to sale, exemplifies precisely this process (see Naylor et al. 2005).

Scholars of Darwin's work and life have been quick to follow his forays outside natural history into political economy, yet, with rare exceptions like Secord (1981, 1985), they have virtually ignored his relations with breeders of domestic animals who were also outside the bounds of natural history at that time. None of the scholars who has written about Darwin's work on animal breeding has asked how the cultural assumptions about human–animal relations in domestication might have been incorporated into Darwin's depiction of ecological relations in "the entangled bank."

Why should we care about the breeders' ideas and practices now? First, Darwin's own position on the place of domestication as a model and means of studying transmutation in nature, yet his disregard of the breeders' views, still raises important questions about the place of human beings in what he described in *Origin* (1859: 426) as "the hidden bond of community of descent" linking all forms of life. Second, the breeders' insistence on their abilities to redesign living organisms according to their own plans remains as deeply entrenched in contemporary understandings of life processes as Darwin's insistence on the randomness of variation and relativity of natural selection. The breeders' position is now the basis of contemporary biotechnology, which has vastly increased the scope and scale of humans' "ecological footprint" (Palumbi 2001) on themselves as well as on their fellow beings, with some strikingly unexpected consequences (e.g., Franklin 2001) that may radically alter our conceptions of wild, tame and human.

Domestication and the "Conditions of Life"

In bringing together this collection of chapters, Molly Mullin and Rebecca Cassidy have asked us to reconsider the concept of "domestication" in anthropology: the ways in which "domestication" has been used to explain relations among and between humans and other creatures; how social, political, or ecological conditions might inform those uses; the relationship of ideas and practices of domestication to oppositions like "wild and tame, "natural" and "human," with which it is commonly associated; the sociocultural, historical, and material dynamics of domestication, in which control over reproduction play a major role and the relationship of domestication to domination more broadly. To me as a social anthropologist, these questions came as a challenge. Although I had a sense of the value of studying domestication in human evolution and archaeology, I had never seriously considered how the subject might be relevant to the ethnographic study of social life in recent centuries.

In retrospect, I attribute my dilemma to a kind of human centrism evident in the commonest understanding of domestication: that domestication has to do with human efforts to tame, adapt, accustom, or force other organisms deemed other- or less human, to human ways of living epitomized in their dwellings in the fullest senses of that term. The evidence here, as in several other cases in this collection, clearly refutes that narrow assumption. In Spitalfields in Darwin's time, pigeon breeding and human breeding were interwoven, in keeping with an avian imagery of generational and biogeographical change in Great Britain, including Darwin's *Origin* and comparable to human–bird relations that scholars have documented worldwide. Human beings, as well as other creatures, live and reproduce in and through the lives and deaths of others; their survivals are entwined. Thus, the study of domestication in any time or place requires us to put human relations in a social framework that includes their mutually transformative relations with other organic beings. Rather than using generic terms like "niche construction," "interaction," "predation," "facilitation," "competition," "symbiosis," "domination," "survival," or "coevolution" in describing common fields of social relations, such an approach, perhaps beginning with codomestication, would encourage more specific documentation and analysis of complex social relations in terms that most anthropologists and other scholars have hitherto limited to the study of human beings (Ingold 1997).

Dwelling in the broadest sense (see Ingold 1996) would be a most fruitful focus, including not only the built environment—from mansions, windscreens, tenements, and cages to the fields, houses, and laboratories of science (Kohler 2002; Shapin 1994: 409–417)—but also "homes without hands" (Wood 1866) that so fascinated people in Darwin's day. Indeed, Turner (2002), a physiologist, goes so far as to argue that these constructions are manifestations of an "external physiology" emergent from processes of mutual adaptation. Lewontin (2000: 67), an evolutionary biologist, argues that "all organisms construct their own environments and that there are no environments without organisms." We must understand them in common social, cultural, historical, and material terms.

Notes

Acknowledgments. This chapter is based on research in archives (including "woven documents") and published primary sources (government publications, newspapers, periodicals, pamphlets, and books): *in London*: at the British Library, Guildhall Library, the T. H. Huxley Scientific Papers, Imperial College/London, State Apartments and Royal Ceremonial Dress Collection at Kensington Palace, Tower Hamlets Local History Library and Archives, the Textiles Division of the Victoria and Albert Museum, Textiles Division; *in the United Kingdom outside London*: the Charles Darwin Archives at the University Library, English Heritage's Darwin Collection at Down House (Downe); *and in the United States*: at Hatcher Library, University of Michigan–Ann Arbor. I am grateful to Molly Mullin and Rebecca Casssidy for having created the Wenner-Gren symposium on "Where the Wild Things Are Now"; to Richard Fox, Laurie Obbink, and Amy Perlow of the Wenner-Gren Foundation for providing the fruitful circumstances of our meetings; and to all my fellow participants for their insights and enthusiasm.

1. According to the U.S. Department of Agriculture's (USDA) *Animal Welfare Report, Fiscal Year 2001,* 1,236,903 animals were used in research in FY 2001, including "dogs, cats, primates, guinea pigs, hamsters, rabbits, sheep, pigs, other farm animals, and other animals" (2002: 31–32), supplied by commercial breeders and exhibitors (licensed and registered). According to the National

Association for Biomedical Research (Topics 2002: 2), this number represents about 4 percent of the total used in research in FY 2001. The roughly 30 million animals in the remaining 96 percent are commercially bred rats, mice, and birds, which are not covered by the USDA's welfare rules. Birds include quails, pheasants, finches, ostriches, doves [pigeons], and parrots, as well as "production birds" like chickens, ducks, and turkeys (Crawford et al. 2004). Three years earlier in FY 1998, 23 million rats, mice, and birds were used as lab animals and the number was expected to increase by one-half in the next three to five years (Trull and Rich 1999: 1463), as indeed it seems to be doing. Commercial breeders now dominate the production of lab animals. Historically, as Rader (2003, this volume) shows, "fancy mice ... were quite literally the 'raw' materials for the creation of laboratory mice." The work of Wood and Orel 2001 suggests that domesticates are the source of most of the other laboratory animals, whether or not they are still under the aegis of the USDA.

2. In fact, Darwin's views changed gradually. The *Notebooks* (e.g. D: 100–104, written in 1838), show that he was concerned about possible limits to variation in domesticates. In zoological nomenclature, Rock Pigeons, domestic pigeons, and feral pigeons are all *C. livia* (Johnston and Janiga 1995: 14–15). Yet feral pigeons "are probably from half again to nearly twice as genically polymorphic as any other birds," including the Rock Pigeons with which they still interbreed (Johnston and Janiga 1995: 37).

3. Feeley-Harnik (2004) includes more data on this case study.

4. According to Johnston and Janiga (1995: 6–8), Rock Pigeons were the first creatures (after dogs around 14,000 B.P.) to enter into domestic relations with humans, probably synanthropically in the Near and Middle East toward the end of the Pleistocene around 12,000 to 10,000 B.P. More recently, Dobney (2002) has suggested that falcons and other birds of prey may have been domesticated around the late Pleistocene–early Holocene (~12,000–10,000 B.P.) in the Middle East.

References

Barrett, Paul H., Peter J. Gautrey, Sandra Herbert, David Kohn, Sydney Smith, eds. 1987. *Charles Darwin's notebooks, 1836–1844: Geology, transmutation of species, metaphysical enquiries.* Cambridge: Cambridge University Press.

Bartley, Mary M. 1992. Darwin and domestication: Studies on inheritance. *Journal of the History of Biology* 25: 307–333.

Brent, Bernard Pierce. 1853. "Time was. ..." *The Cottage Gardener* 11: 35–38.

Burkhardt, Frederick H. et al., eds. 1985–2002. *The correspondence of Charles Darwin*, vols. 1–13 (1821–65). Cambridge: Cambridge University Press.

Cassidy, Rebecca. 2002. *The sport of kings: Kinship, class and thoroughbred breeding in Newmarket.* Cambridge: Cambridge University Press.

Cooke, Arthur O. 1920. *A book of dovecotes.* London: T. N. Foulis.

Corres. *See* Burkhardt et al. 1985–2002.

Crawford, Richard L., D'Anna Jensen, Heidi Erickson, and Tim Allen. 2004. *Housing, husbandry, care & welfare of selected birds (quail, pheasant, finches, ostrich, dove, parrot & others), AVIC resource series, 26—February 2004.* Washington, DC: Animal Welfare Information Center for the U.S. Department of Agriculture. Available at: http://www.nal.usda.gov/awic/pubs/Birds/birds.htm.

Cuvier, Georges. 1822–23. "Domestication," translated by Robert Jameson. *Edinburgh Philosophical Journal*, vols. 6–8.

Darwin, Charles. 1831–36. *Diary of the voyage of H. M. S. Beagle*, edited by Nora Barlow. Reprinted in *The works of Charles Darwin*, vol. 1, edited by Paul H. Barrett and Richard B. Freeman. New York: New York University Press, 1987.

——. 1839. *Journal of researches into the geology and natural history of the various countries visited by H.M.S. Beagle.* Reprinted in *The works of Charles Darwin*, vols. 2–3, edited by Paul H. Barrett and Richard B. Freeman. New York: New York University Press, 1987.

——. 1842, 1844. *The foundations of the* Origin of Species. *Two essays written in 1842 and 1844*, edited by Francis Darwin. Reprinted in *The works of Charles Darwin*, vol. 10, edited by Paul H. Barrett and Richard B. Freeman. New York: New York University Press, 1987.

——. 1859. *On the origin of species by means of natural selection, or the preservation of favoured species in the struggle for life.* Facsimile edition, introduced by Ernst Mayr. Cambridge, MA: Harvard University Press, 1964.

——. 1868. *The variation of animals and plants under domestication*, 2 vols. London: John Murray.

——. 1870–71. *The descent of man, and selection in relation to sex*, 2 vols. London: John Murray.

Darwin, Charles, and Alfred Russel Wallace. 1859. On the tendency of species to form varieties [Darwin]; and On the perpetuation of varieties and species by natural means of selection [Wallace]. *Journal of the Proceedings of the Linnean Society (Zoology)* 3 (1859): 45–62.

Diary. *See* Darwin 1831–36.

Dickens, Charles. 1998 [1864–65]. *Our mutual friend,* edited with an introduction and notes by Michael Cotsell. Oxford: Oxford University Press.

Dixon, E. S. 1851. *The dovecote and the aviary: Being sketches of the natural history of pigeons and other domestic birds in a captive state, with hints for their management.* London: W. S. Orr.

Dobney, Keith. 2002. Flying a kite at the end of the ice age: The possible significance of raptor remains from proto- and early Neolithic sites of the Middle East. In *Archaeozoology of the Near East,* vol. 5, edited by H. Buitenhuis, A. M. Choyke, M. Mashkour, and A. H. Al-Shiyab, 74–84. Groningen, the Netherlands: ARC-Publicaties 62.

Domestic. 1841. The domestic pigeon. *The Penny Magazine* 10: 361–363.

Ellen, Roy, and Katsuyoushi Fukui, eds. 1996. *Redefining nature: Ecology, culture, and domestication.* Oxford: Berg.

Essays. *See* Darwin 1842, 1844; and Barrett et al. 1987.

Feeley-Harnik, Gillian. 2004. The geography of descent. *Proceedings of the British Academy* 125: 311–364.

——. n.d. Tame Birds: Sexuality and Domestication. MS in the author's possession.

Feld, Steven. 1987. *Sound and sentiment: Birds, weeping, poetics, and song in Kaluli expression.* 2nd edition. Philadelphia: University of Pennsylvania Press.

Flannery, Kent V., Joyce Marcus, and Robert G. Reynolds. 1989. *The flocks of the Wamani: A study of llama herders on the punas of Ayachucho, Peru.* New York: Academic Press.

Franklin, Sarah. 2001. Sheep watching. *Anthropology Today* 17: 3–9.

Friedrich, Paul. 1997. An avian and aphrodisian reading of the *Odyssey. American Anthropologist* 99 (2): 306–320.

Fulton, Robert. 1874–76. *The illustrated book of pigeons. With standards for judging,* edited by Lewis Wright, illustrated by J. W. Ludlow. London: Cassell Petter and Galpin.

Galton, Francis. 1865. The first steps towards the domestication of animals. *Transactions of the Ethnological Society of London* 3: 122–138.

——. 1883 [1865]. "The First Steps towards the Domestication of Animals," with new introductory and concluding paragraphs as "Domestication of Animals." In *Inquiries into human faculty and its development,* 173–194. London: Dent.

Haeckel, Ernst. 1866. *Generelle Morphologie der Organismen. Allgemeine Grundzüge der organischen Formen-Wissenschaft, mechanisch begründet durch die von Charles Darwin reformirte Descendenz-Theorie.* 2 vols. Berlin: G. Reimer.

Holloway, Simon, comp. 1996. *The historical atlas of breeding birds in Britain and Ireland: 1875–1900*. London: T. and A. D. Poyser for the British Trust for Ornithology.

Huxley, Thomas H. 1862. *On our knowledge of the causes of the phenomena of organic nature*. London: Robert Hardwicke.

Ingold, Tim. 1994. From trust to domination: An alternative history of human–animal relations. In *Animals and human society: Changing perspectives*, edited by Aubrey Manning and James Serpell, 1–22. London: Routledge.

———. 1996. Growing plants and raising animals: An anthropological perspective on domestication. In *The origins and spread of agriculture and pastoralism*, edited by David R. Harris, 12–24. London: University College London Press.

———. 1997. Life beyond the edge of nature? Or, the mirage of society. In *The mark of the social*, edited by J. B. Greenwood, 231–252. Lanham, MD: Rowman and Littlefield.

Johnston, Richard F., and Marián Janiga. 1995. *Feral pigeons*. New York: Oxford University Press.

Kohler, Robert E. 2002. *Landscapes and labscapes: Exploring the lab–field border in biology*. Chicago: University of Chicago Press.

Leach, Helen M. 2003. Human domestication reconsidered. *Current Anthropology* 44: 349–368.

Lewontin, Richard. 2000. *The triple helix: Gene, organism and environment*. Cambridge, MA: Harvard University Press.

Lucas, J. (1886), *The pleasures of a pigeon-fancier*. London: Sampson Low, Marston, Searle, and Rivington.

Lyell, Charles. 1830–33. *Principles of geology, being an attempt to explain the former changes of the earth's surface, by reference to causes now in operation*. 3 vols. Facsimile of the first edition, introduced by Martin J. S. Rudwick. Chicago: University of Chicago Press, 1991.

Manchée, William. 1918. George Dorée, citizen and weaver of London, b. 1844, d. 1916, the maker of King Edward VII's coronation velvet. *Proceedings of the Huguenot Society of London* 11: 419–425.

Mayhew, Henry. 1849. Labour and the poor. The metropolitan districts. Letter 2 ["Bethnal-green … Spitalfields weavers"]. *The Morning Chronicle*, 23 October: 5.

Morgan, Lewis Henry. 1881. *Houses and house-life of the American aborigines*. Reprinted, with a foreword by William Longacre. Salt Lake City: University of Utah Press, 2003.

Mullin, Molly. 1999. Mirrors and windows: Sociocultural perspectives on human–animal relationships. *Annual Review of Anthropology* 28: 201–224.

Naylor, Rosamond, Henning Steinfeld, Walter Falcon, James Galloway, Vaclav Smil, Eric Bradford, Jackie Alder, and Harold Mooney. 2005. Loosing the links between livestock and land. *Science* 310 (9 December): 1621–1622.

Newton, A. 1858 [1853–60]. Ornithology. In *Encyclopaedia Britannica*, 8th ed., vol. 16: 792–794.

Notebooks. *See* Barrett et al. 1987.

Our Mutual Friend. 1864–65. *See* Dickens 1998/1864–65.

Palumbi, Stephen R. 2001. *The evolution explosion: How humans cause rapid evolutionary change.* New York: W. W. Norton.

Parliamentary Papers (P. P.). 1823. "Minutes of Evidence relative to the wages of the Silk Manufacturers"; P. P. 1823 sess. 1 HL 86 xiii 1.

——. 1831–32. "Report from the Select Committee on the Silk Trade"; P. P. 1831–32 sess. 1 (678) xix 1.

P. P.–Fletcher. 1840. "Hand-loom weavers: Reports … Part IV … by J. Fletcher on the Midland Districts of England"; xxiv (217) 1.

P. P.–Kay. 1837. "Report from Dr. Kay to the Poor Law Commissioners on the subject of Distress in Spitalfields"; P. P. 1837 (376.) li 191.

P. P.–Hickson. 1840. "Hand-loom weavers: Reports … by W. E. Hickson on the condition of the hand-loom weavers"; xxiv (636) 639.

P. P.–Mitchell. 1840. "Hand-Loom weavers: Reports from assistant hand-loom weavers commissioners … Part II … by J. Mitchell on the East of England"; P. P. 1840 xxiii (43.XI) 53.

Rader, Karen. 2003. *Making mice: Standardizing animals for American biomedical research.* Princeton: Princeton University Press.

Reina, Ruben E., and Kenneth M. Kensinger, eds. 1991. *The gift of birds: Featherwork of Native South American peoples.* Philadelphia: University Museum of Archaeology and Anthropology.

Russell, Nerissa. 2002. The wild side of animal domestication. *Society and Animals* 10: 285–302.

Secord, James A. 1981. Nature's fancy: Charles Darwin and the breeding of pigeons. *Isis* 72: 163–186.

——. 1985. Darwin and the Breeders. In *The Darwinian heritage,* edited by David Kohn, 519–542. Princeton: Princeton University Press.

Shapin, Steven. 1994. *A social history of truth: Civility and science in seventeenth-century England.* Chicago: University of Chicago Press.

Stauffer, Robert C. 1957. Haeckel, Darwin, and ecology. *The Quarterly Review of Biology* 32: 138–144.

Sulloway, Frank. 1982a. Darwin and his finches: The evolution of a legend. *Journal of the History of Biology* 15: 1–53.

——. 1982b. Darwin's conversion: The *Beagle* voyage and its aftermath. *Journal of the History of Biology* 15: 325–396.

Taine, Hippolyte. n.d. [ca. 1871]. *Notes sur l'Angleterre (1860–1870)*. Paris: Simon Raçon.

Thirsk, Joan. 1992. English rural communities: Structures, regularities, and change in the sixteenth and seventeenth centuries. In *The English rural community: Image and analysis,* edited by Brian Short, 44–61. Cambridge: Cambridge University Press.

Terrell, John Edward, John P. Hart, Sibel Barut, Nicoletta Cellinese, Antonio Curet, Tim Denham, Chapurukha M. Kusinba, Kyle Latinis, Rahul Oka, Joel Palka, Mary E. D. Pohl, Kevin O. Pope, Patrick Ryan Williams, Helen Haines, and John E. Staller. 2003. Domesticated landscapes: The subsistence ecology of plant and animal domestication. *Journal of Archaeological Method and Theory* 10: 323–368.

Thomas, Keith. 1983. *Man and the natural world: Changing attitudes in England, 1500–1800.* London: Allen Lane, Penguin.

Thrift, Nigel, and Peter Williams, eds. 1987. *Class and space: The making of urban society.* London: Routledge and Kegan Paul.

Topics. 2002. Topics in animal care, November–December, 2002. Animal Resources Center, University of Chicago. Available at: http://arc.bsd.uchicago.edu/node02.html.

Trull, Frankie L., and Barbara A. Rich. 1999. More regulation of rodents. *Science* 284 (May 28): 1463.

Turner, J. Scott. 2000. *The extended organism: The physiology of animal-built structures.* Cambridge, MA: Harvard University Press.

U.S. Department of Agriculture (USDA). 2002. *Animal welfare report, fiscal year 2001.* Washington, DC: Animal and Plant Health Inspection Service for the USDA. Available at: http://www.aphis.usda.gov/ac/awrep2001.pdf.

Warner, Frank. ca. 1921. *The silk industry of the United Kingdom: Its origin and development.* London: Drane's.

Wilson, Peter J. 1988. *The domestication of the human species.* New Haven, CT: Yale University Press.

Wood, James G. 1866. *Homes without hands. Being a description of the habitations of animals, classed according to their principle of construction.* London: Longmans, Green.

Wood, Roger J., and Vítězslav Orel. 2001. *Genetic prehistory in selective breeding: A prelude to Mendel.* Oxford: Oxford University Press.

Wright, Lewis. 1879. *The practical pigeon keeper.* London: Cassell.

Yarrell, William 1837–43. The Rock Dove [Part 17, 2 March 1840]. In *A history of British birds,* 2: 259–266. 3 vols. London: Van Voorst.

seven

The Metaphor of Domestication in Genetics

Karen Rader

What is domestication and how can and should we understand it? My focus is on one very specific aspect of this inquiry—namely, domestication of nonhuman animals in and for the laboratory—because I am interested in exploring implications that the continued use of both the metaphors and the practices of domestication have had for relations among all contemporary scientific animals (human and nonhuman). In part, my analysis addresses a recurring historical argument about what is or is not different between the so-called "pregenomic" and the "postgenomic" era of biological and agricultural research—although, as Human Genome Project head Francis Collins noted in 2002, those terms themselves have dubious meaning.[1] But in so doing, it also transforms this question from a merely empirical one about history to one of political anthropology: Why should domestication matter to us, as analysts and citizens of a contemporary world in which biology and social policy are co-constructed spheres? Is domestication a useful resource with which to consider the present and future of animal–human relations in our technoscientific world, as well as in the past? Do new forms of biological control in the laboratory require us to redefine what we mean by domestication—or rather, a la Donna Haraway's categorical "implosions" of machine and natural life forms, and Paul Rabinow's "biosociality"—or should the concept be abandoned because it relies on an analytically dubious nature–culture dichotomy that does no longer—or perhaps never did—apply (Haraway 1997; Rabinow 2003)?

Among anthropologists, "domestication" stands out as a traditional concern since this discipline began. But as Rebecca Cassidy points out (Introduction this volume), domestication as an object of inquiry remains firmly anchored in history. By contrast, for those whose work concerns making and reporting on U.S. policymaking around genetically engineered (GE) organisms, the stakes of "what is domestication?" are quite present and quite clear. In the United States and internationally, genetic engineering practices are often defended as "just another form of domestication"—that is, accelerated but continuous with other kinds of conventional selective breeding (particularly, in agriculture) that have gone on safely in human societies for millennia, or at the very least, in laboratories for centuries (Trudinger 2003; see also Pew Initiative on Food and Biotechnology 2002a, 2002b). We can keep doing it, the argument goes, because it is nothing more than we have always already been doing.

There are many good analytic reasons to question the notion that the future course of domestication is inevitable. Helen Leach warns against such human centrism (Leach 2003), and Haraway cautions that we must be politically skeptical of an "attitude towards history [that often] characterizes those who live in the timescape of the technopresent" and tend to "describe everything as new, as revolutionary, as future-oriented solutions to the problems of the past" (Haraway 2003: 297). Archaeological anthropologists describe such problems with the very idea of a "domestic animal." Nerissa Russell argues that precisely because domestication "provides an important tool in power negotiations among humans as well as between humans and animals" (2002: 286), analysts must self-consciously strive to understand it as a category—which at once constructs and crosses boundaries between human and animal, nature and culture—as well as a process. But where, then, does all this reflexivity get us regarding the meaning and uses of domestication in the historical and contemporary genetic engineering of animals?

Photographer Yann Arthus-Bertrand's work *Good Breeding* (1999) points in a productive direction for addressing this question, because aesthetically, it starkly evokes the convergence of human and animal domestication. Arthus-Bertrand's photographs of prizewinning breeds of livestock and their owners—set against simple canvas backdrops at agricultural shows in England, France, Italy, and Argentina—have caused more than one reviewer to remark on the correspondences: "The animals look strangely like their owners (or is it the other way around?)" (reviewer, September 1999, http://www.amazon.com, accessed on December 19, 2003; see also Becker 2000). Conceptually, then, these representations also domesticate the livestock breeders Arthus-Bertrand depicts, thereby inviting viewers to ask what amount to profound ontological questions about history and agency—such as: What is it that these pictures witness—historical coincidence or evolutionary process —and who caused it and why? And finally, and perhaps most importantly, *why*, when we look at these pictures, might we envision these representations of humans and animals similarly?

Given the material and social conditions of its production, Arthus-Bertrand's work initiates simultaneously a bioethical inquiry about the artist's domesticating practices and a biopolitical assessment of domestication. Arthus-Bertrand himself must sustain close relationships with livestock breeders—a notoriously iconoclastic lot, practically close to their animals for their own (human) livelihood but often unsentimental about them—to have access to the subject of his work. Arthus-Bertrand creates artificial scenarios (lighting, backdrops) at unnatural settings (cattle and dog shows) to capture his objects of interest where they "naturally" gather. In each frame, what is absent is as important as what is present for understanding what it might mean to claim domestication (of the artist, as well as his subjects), and its many convergences, as a useful metaphor. By conflating the domesticated and the domesticator—that is, the source and the object of power relations between his human and animal subjects—Arthus-Bertrand's work collapses the metaphorical discourse of domestication, an act that is itself is an intervention in the process of biopolitical knowledge making.

Arthus-Bertrand is certainly not the first person to make such an intervention: the history of efforts to genetically engineer laboratory mammals demonstrates that the metaphor of domestication—and its collapse—has proven very useful scientifically for uniting disparate domains of practice and history in animal–human relations. Beyond well-known attempts by Darwin and Galton (1865a, 1865b, 1883) to apply the concepts and/or methods of artificial selection to understand and/or influence the biology of natural selection, 20th–century geneticists have a long, relatively unknown and unappreciated history of relying on both the practices and the metaphors of domestication in their work. This chapter briefly details this history to suggest how resources brought to bear to understand the practices of laboratory domestication can be also relevant for understanding the evolution of the metaphor of domestication more broadly. Following Nancy Stepan and others, I presume that the choice of particular biological metaphors has "social and moral consequences, in addition to intellectual ones" (Stepan 1986: 275). In the case of art as well as science, similarity is not something one finds but something one must establish—and thus meaning is the product of the relation between two parts of a metaphor, but so too is the power of that meaning. Ultimately, then, I suggest that contemporary scientific work with GE laboratory mammals has much in common with past practices of domestication: it, too, deploys new technologies at the interface of lay and scientific cultures to collapse and reinscribe yet again Western cultural boundaries between humans and animals and between nature and nonnature. In so doing, this work offers us a future of newly naturalized and unstoppable marketplace inevitabilities. But I argue that, instead of seeing historical determinism in domestication, we might see in the elements of its relationality a possible point of intervention in contemporary debates over the future of genetic engineering and biopolitics more broadly.

Fanciers, Geneticists, and Other Humans: Early Domestication of the Mouse

Gregor Mendel—the Moravian monk most often credited with founding genetics—was himself a domestic breeder of peas and bees, so perhaps it is not surprising that when Mendel's work was rediscovered in 1900, U.S. breeders and agriculturalists were among the first to embrace it as a true science of inheritance (Paul and Kimmelman 1988; see also Kimmelman 2003). In turn, early geneticists recognized that specific variants of highly inbred animal populations—in current scientific parlance, "mutants"—could be used effectively as tools to sort many heritable features of organisms into biologically identifiable processes. In practice, however, such efforts were just as frequently taken on by so-called "amateurs": for example, German high school teacher Hans Duncker combined his rudimentary knowledge of genetic science with the bird-breeding expertise (ability to select for color and song) of fancier and shopkeeper Karl Reich in a quest to create what biologist Tim Birkhead claims was the first "genetically engineered" animal: a red canary (Birkhead 2003).

But the mutants used most productively in early academic genetic research did not come initially from highly prized cultivars; they were pests such as the fruit fly, *Drosophilia*. As historian Robert Kohler has shown, Columbia University zoologist T. H. Morgan and his "boys" (then–graduate students Calvin Bridges and Arthur Sturtevant) exploited this organism's proximity to humans and its biological capacity as a "breeder reactor"; Drosophila bred fast and generated copious mutants, which enabled Morgan's team to construct the first animal genetic map of its four salivary chromosomes in 1921 (Kohler 1995). For Kohler, as well as for his historical subjects, such laboratory domestication of the fly was continuous with its biological evolution and domestication more broadly and it created the kind of mutual dependency between the human and animal actors that Stephen Budiansky argues is the hallmark of domestication itself (Budiansky 1992). As Kohler writes, "when fruit flies crossed the threshold of the experimental laboratory, they crossed from one ecosystem to another quite different one, with different rules of selection and survival. ... Once in the lab, *Drosophila* ... revealed an unexpected and very remarkable capacity for experimental heredity and genetics which soon made it and its human symbionts famous" (Kohler 1995: 19). Soon after technological processes—such a X-rays and chemical mutagenesis—were developed to create even more fly mutants, which sealed Drosophila's fate in the development of classical genetics, and thereafter, in biological teaching laboratories (Muller 1928).

Mice have also been "hangers-on" to human culture for thousands of years, so their cultural, as well as biological, identity derives first and foremost from that relation. Taxonomically, mice are the smallest members of the order *Rodentia,* or "gnawers," and their evolutionary appearance dates to the Eocene epoch, 54 million years ago. But the present scientific name of the common house mouse, from which inbred laboratory mice are descended, has little to do with the creature's appearance. In *Mus musculus,* the Latin *mus* derives from an ancient Sanskrit word, *musha,* meaning "thief," which suggests that the mouse and its predatory feeding habits were familiar to cultures in Asia dating back before 4000 B.C.E. But regardless of when exactly they first started to associate with humans, mice quickly became successful commensals. House mice were common in the earliest farming villages of present-day northern Iran, as well as in areas surrounding the ancient Mediterranean. Some ancient civilizations recorded virtual plagues of mice—sometimes accompanied by disease—and modern biologists speculate that this prolific species increased its geographical dispersion by accompanying many early human migrations to present-day Europe and Africa (Moulton 1901). From there, *"Mus musculus* proper shared with the European his recent conquest of the globe" as ocean-vessel stowaways to all habited regions of the Asiatic seacoast and to the Americas (Keeler 1931: chs. 1–2, 1932; cf. Beckman 1974; de Gubermatis 1968/1872; Grohmann 1862; Malriey 1987).

Although the internationalization of the house mouse is a relatively recent phenomenon as measured by evolutionary time, ancient cultural traditions express many mouse mythologies and narratives. Some such legends are pejorative. For example, Aelieanus (ca. C.E. 100) of Lower

Egypt speculated disgustedly that mice developed from raindrops because they were so plentiful in that area. More recently, anthropologists have suggested that the Egyptians' hatred of mice accounts in large part for their well-known deification of the cat. Yet other stories and cultural images have cast the mouse in a positive light. More than 1,000 years before Christ, Homeric legend reported a cult of the mouse–god *Apollo Smintheus,* whose popularity reached its height around the time of Alexander the Great. White mice, because of their relative rarity and their associations with purity, were thought to forecast prosperity in a home. Beliefs in the healing powers of these mice originated with the Apollo worshippers and persisted among medieval scholars such as Hildegard of Bingen. Other colors of mice were also venerated; temples in Troas, for example, held marble sanctuaries overflowing with gray mice raised at public expense (Keeler 1932).

Many aspects of the mouse's ancient cultural legacy persisted in the United States through the early decades of the 20th century, but by the 1930s, these representations existed simultaneously with positive cultural depictions.[2] The modern heirs to the ancients' negative portrayals of mice included various accounts—both scientific and folkloric—of mice as harbingers of disease and "spookers" of women. At the same time, Walt Disney's personified Mickey Mouse made his public debut in 1928 and proved phenomenally popular.[3] Also in 1929, the popular magazine *Nature* ran an article entitled "White Mice," which fancifully described the daily activities of the author's real pet mice. The author gave a particularly pointed and modern spin to Egyptian "pure white mouse" myth: "If cleanliness is next to godliness, as the soap advertisements say, then Plato was wrong and our animals do have souls" (Johnson 1929).

For genetics, however, the most significant early 20th-century human activity explicitly centered around mice was a hobby called "mouse fancying." The exact origins of mouse fancying are obscure, although textual sources (believed to be early breeding manuals) indicate that the collecting and developing unique strains of mice in captivity dates as far back as 17th-century Japan (Grüneberg 1943). But the formation of many local and national mouse fancier organizations in the early 1920s indicates that mouse fancying clearly enjoyed increased popularity in the United States and Britain beginning in the early 20th century. Fanciers who belonged to the American Mouse Fanciers club, and its many British counterparts, selected for certain "standard" physical features and preserved the specimens that exhibited them ("Mice Beautiful" 1937). As described by a 1930s popular magazine article, fanciers thought "the perfect mouse should be seven to eight inches long from nose-tip to tail-tip, the tail being about the same length as the body and tapering to an end like a whiplash" ("The English Craze for Mice" 1937: 19). Fanciers most often kept these mice as pets and would travel with them to local or national "mouse shows," which awarded small cash awards to the owners of visually unusual and interesting specimens. Other mouse-breeding enterprises had more lucrative commercial interests in mind. In 1930s England, for example, mouse breeders could cash in on the demand for full-length women's coats made of mouse skins, which took 400 skins and sold for $350 retail ("Mouse Show" 1937).

One fancier, in particular, was of significance for the development of laboratory mice: Abbie Lathrop, who ran the Granby Mouse Farm in Granby, Massachusetts. Lathrop founded this institution around 1903 as an alternative to her failing poultry business. Mice and rats, for sale as pets, provided an inherently quicker turnover, and Lathrop probably believed the growing community of fanciers in the New England area would be her main market. But instead of receiving requests for a few mice of exotic coat color from mouse fanciers, she soon began to get large orders for mice from scientific research institutions and medical schools. Lathrop's farm quickly became the East's largest supplier of mice for research in the first two decades of the 20th century. She took orders from laboratories all along the East coast and from as far West as St. Louis.[4]

By 1913, the Granby Mouse Farm had become such a local curiosity that a Massachusetts newspaper devoted a feature article to it. The details of care taking provided in the reporter's account reveal clearly that Lathrop's mouse breeding for research was a large and resource-intensive undertaking, requiring extensive practical knowledge of proper mouse husbandry. Her stocks had gradually increased to 10,000 since her humble beginnings with "a single pair of waltzing mice which she got from this city." Lathrop housed her mice in wooden boxes, with straw as a bedding material, and because cleaning the cages had become too much work for her alone, she periodically hired town children at seven cents an hour for this purpose. The mice were fed a diet of crackers and oats, and Lathrop reported going through 12 and 1/2 barrels of crackers and a ton and a half of oats each month. Furthermore, the cages were given fresh water daily "in little jars which are first boiled as protection against disease germs." Lathrop even appears to have experimented with a primitive water bottle device, from which "a thirsty mouse has only to stand on his hind legs to quaff a cooling drink" (*Springfield Sunday Republican* 1913).

Lathrop's fancy mice, then, can rightly be called the "raw materials" for the creation of laboratory mice, and the boundaries between the "field" and "amateur" knowledge making of fanciers, and the "laboratory" "professional" science of genetics remained porous for several decades. Fanciers learned from and exploited their relationship with scientists, and vice versa. For example, Harvard zoology professor W. E. Castle himself attended mouse fancy shows, and encouraged his students to do the same—some even acted as judges. Also, Lathrop herself was interested in science and worked with University of Pennsylvania pathology professor Leo Loeb to breed and analyze patterns of tumor inheritance in several strains (most importantly, one that fanciers had named "silver fawn" for its coat color, but geneticists renamed *dba* as an abbreviation for its coat color genes: dilute, brown, and nonagouti; Rader 2004: ch. 1). One Castle student in particular—C. C. Little—sought out Lathrop's particular variants and the use of these materials shaped his own genetic research, which aimed to make Mendelian sense of mouse coat and eye color inheritance as well as mammalian cancers. Little ultimately translated his vision for the role of inbred animals in research into the Jackson Laboratory, a Bar Harbor, Maine research institute and mouse supplier founded in 1929 and still going—stronger than ever—today (cf. Rader 1999). Another—Freddy Carnochan—cofounded a commercial animal breeding farm called Carworth Farms, and continued to work closely with card-carrying mouse geneticists—especially, L. C. Dunn of Columbia—to identify and develop new mutant stocks.

Some early mouse geneticists, like Castle student Clyde Keeler, self-consciously used the metaphor of domestication—especially the idea that laboratory domestication represented a process continuous with mus musculus' evolution as a human symbiont—to argue that there was no important boundary between past and present practices. For a scientific audience, Keeler wrote *The Laboratory Mouse: Origin, Heredity, and Culture* (1931), a handbook cum homage to his favored laboratory creature. It aims to comprehensively collect "literature upon the house mouse, its origins, history, distribution, development, the nature of its variations, the hereditary transmission of its varietal characters, as well as methods of rearing it suitable for the needs of laboratories" to "present it in a useable form." But what counts as "useable form" amounts to a kind of claiming of past domestication efforts in the name of genetics. The book, for example, provides a list of the most important Mendelian unit characters in mice—the first being dominant spotting in 1100 B.C.E. and the last being George Snell's dwarf mutant in 1929—to show that out of 18 then extant, eight had been recorded since 1900. Likewise, in another article for the popular *Scientific Monthly,* Keeler led his readers on a journey "In Quest of Apollo's Sacred White Mice," only to conclude that although much of the mouse's ancient history lies buried "in the religious auguries of Babylon and Troy," "we may say definitively that Apollo's mice were albinos of the species *mus musculus* and that our laboratory mice are probably descended from the temples of Apollo. [This is] the longest heredity of a simple variation of which we have a written record" (Keeler 1931: 51, 53, respectively).

At the same time, other genetic scientists mobilized existing cultural boundaries between humans and animals, as well as nature and culture, to advance their own domestication practices in the laboratory. In 1935, for example, Little penned a *Scientific American* article he titled "A New Deal for Mice." First, he juxtaposed gains the mouse had made in science during the last decade with a prevailing cultural stigma of the animal: "Do you like mice? Of course you don't. 'Useless vermin,' 'disgusting little beasts,' or something worse is what you are likely to think as you physically or mentally climb a chair." Then against this background, Little cast himself as "attorney for the defense," and argued that through their involvement with science, mice had been positively transformed. Inbred laboratory mice—as opposed to their "not very convenient" wild mice relatives—"provided a particular service" to both science and to humanity. Little invited his lay readers to visit the *domus* through which this became possible: the Jackson Lab's "mouse house" or, in another more Progressive description, one of the "mouse laboratory 'cities' with its cleanliness, orderly arrangement, and activity." Such arrangements testified that "thoroughbred" mice (a concept Little acknowledged some people would find "amusing") had become "an integral part of man's helpers." "Under these circumstances," Little concluded, "perhaps mankind will accept and develop his relationships with mice in a different light" (Little 1937).

In all these cases, however—some rhetorically self-conscious, and others practically strategic—domestication functioned as an active, relational, if sometimes contradictory, meaning-making metaphor that united and ultimately naturalized the coexistence of diverse domains of knowledge making and history in mouse–human relations. Mouse fanciers routinized the activity of mouse breeding in captivity well before scientists became interested in this animal—and in so doing, established traditional husbandry assumptions while lowering the practical thresholds to mouse use in the laboratory. Fanciers provided genetic scientists with both a unique mammalian material resource and a broader practical context in which the controlled breeding of animals for human ends (beyond food) was an accepted cultural activity. But scientists' active collaboration with this group of animal producers and pet owners also highlights the kind of ambiguous and fluid boundaries that separated domestic fancy animals (even nonagricultural ones) from humans.

The cultural turn to "mice-as-pets" that the mouse fancy represented could potentially have resulted in increased emotional attachment to the species (Midgley 1983: chs. 9–10)—and resistance to their use in laboratories. In fact, this was not the trend: mouse fanciers were more interested in what the new science of genetics could do for them than in viewing it critically. Fanciers sought to use genetic knowledge to understand their own breeding process and get a leg up on the competition (other commercial breeders and/or competitors at mouse fancy shows). In turn, thanks to the existence of fancy mice, mice did not need to be trapped messily from one's home or field to be obtained for research. Instead, mice could be ordered from a breeder, making contact with them in their "natural" state unnecessary—as shown by the highly stylized pictures of fancy mouse variants from a Jackson Laboratory catalog, circa 1950.

Many of these early domesticating projects of mammalian geneticists were linked to making their discipline more relevant to larger U.S. Progressive goals. For example, both George Snell and L. C. Dunn have said that their ultimate decisions to do such genetics with mice rather than with flies hinged on a belief that such work would be (in Dunn's words) a more "novel" and socially useful contribution to the field. Little became preoccupied with mice while listening to Castle's undergraduate genetics lectures and appears to have chosen to do murine genetics—instead of the dog coat color genetics he first intended to pursue—because such mouse research would ultimately have relevance to larger biomedical problems of public health (Little 1916; cf. Provine 1989: 60). Case studies by Barbara Kimmelman (corn; 2003) and Judith Johns Schloegel (paramecium; 2006) show that the mouse geneticists were not alone in embracing this goal. Philip J. Pauly has even gone so far as to argue that the impulse "to culture"—understood in its 19th-century meaning as cultivation—marks a significant feature of early 20th-century biology, which was defined (except in universities) by "an ongoing effort on the part of scientists in the United states to 'culture' the Western hemisphere and its organisms—to influence the distribution, reproduction, and growth of plants, animals, and humans, and to improve them" (2000). It is difficult, if not impossible, to evaluate this grand narrative at this moment in the historiography of biology, because, to generate anything resembling the panoramic view to which Pauly aspires, we need more detailed studies of how the practices of U.S. biology constituted and were themselves constituted by the practices of U.S. politics and culture. But the important point here is this: the metaphor of domestication was one about whose relevance early mammalian geneticists and fanciers agreed. In turn, this metaphor was itself a potent resource through which early mouse breeders achieved knowledge-making power, within the laboratory as well as outside of it.

Human–Animal Relations in the Domestication of Genetic Science

Decades later, a vast enterprise of scientific breeding of animals for research in the laboratory now exists on top of an even vaster enterprise of scientifically breeding animals for the purpose of companionship or recreation—and the metaphors and practices of domestication remain powerful tools that actors in these technoscientific worlds use to make sense of their work. By best estimates (which are always old and methodologically virtually guaranteed to be low), the numbers of research mice bred and sold alone are staggering: In 1965, a total of nearly 37 million mice were consumed in U.S. laboratories, and by 1984, that figure had risen to an estimated 45 million—or 63 percent of the total number of (counted) animals used by U.S. scientists (Rowan 1984: ch. 5; cf. National Institutes of Health [NIH] 1990).[5] The primary producers of laboratory animals are now commercial or industrial labs (such as Charles River Laboratories), but various academic scientific institutions (such as the Jackson Laboratory and the NIH) retain a significant market share (Rader 2004).

Alongside the expansion of laboratory animal domestication, animal fancying has persisted; biologist Tim Birkhead estimates: "in each case, starting with a single species, humans have created more than three hundred breeds of domestic pigeons, over one hundred dogs, dozens of breeds of cats, mice, sheep, pigs, and cattle—and some seventy canary breeds" (2003: 87). Animal fancying might even to be said to be becoming part of the 21st-century cultural mainstream again. Some examples: the last Rodent Fancy show held in New York was reported in the front page of the *New York Times* (1996), and the Westminster Kennel Club Annual Dog Show is now televised (on CBS, a major network), and was itself the subject of parody in the successful 2002 Christopher Guest film, *Best in Show*. Indeed, animal breeding for the laboratory and for the fancy also remains closely connected in the scientific imagination: mammals such as dogs, rats, and mice were chosen by the NIH for model organism genome projects in part because geneticists understood that these long-inbred domesticates represented invaluable controlled material resources for the study of inheritance. The completion of the so-called "mouse genome project" was announced in December 2002, followed by the publications of the rat genome in April 2004, and the dog genome in July 2004 ("Model Organisms" at www.genome.gov, 2002).

At the same time, the boundary between human and animal domestication—once practically reconstituted when the Western world acknowledged the horrors of Nazi eugenic practices—now seems to have been made blurry again by the reproductive processes employed on domesticated animals in the laboratory. Genetic and reproductive technologies designed to further human ends—especially with regard to reproduction—have a long history of borrowing freely from knowledge obtained through the controlled breeding of domesticated animals— and vice versa (Pincus and Saunders 1939; Franklin 1997, 2003; cf. Thompson 2001; Ginsburg and Rapp 1995). But now, even when human reproduction is not an aim of a particular research program involving a GE animal, it is frequently invoked—by scientists and pundits alike—in conversations about the application of results. For example, Ian Wilmut, Dolly the sheep's scientist–creator, has noted that the main objective of the controlled breeding program that resulted in her birth was not achieving reproductive cloning itself but, rather, devising a strategy for genetic manipulation of farm animals. Dolly originated in a project to create a sheep ("Tracy") that would exert in her milk large quantities of AAT (alpha-1-antitrypsin), an enzyme used to treat human lung disorders and formerly only available through its extraction from human blood plasma (Wilmut, Campbell, and Tudge 2000: 20–21, 30–32). But this animal "pharming"—and its own implications for reframing political and social relationships between humans and animals, as well as for the health care economy—was not what captured the U.S. cultural and scientific imagination; rather, it was the possibility that cloning technology could be applied to human reproduction. Dolly died in February 2003 (ironically, of a lung infection), almost two years to the day after the decoding of the Human Genome Project was announced. The moral of Dolly, Princeton biologist Lee Silver argued, is that although the technology is currently inadequate, human reproductive cloning through nuclear transfer between two human cells "is a story that will come true. To believe otherwise is to misunderstand the power of the marketplace, and the power of individual desires to reach very specific individual reproductive goals" (Silver 2001: 63; cf. Holt and Williams 2003; Strathern 1992).

Dolly's appearance, then, marked a significant step in the process that Sarah Franklin has described as "the defamiliarization of what it means to do biology or be biological" (Franklin 2000: 203). Silver and other advocates of newer forms of genetic engineering actively seek to obscure the boundaries between human and animal domestication, at the same time that they renaturalize domestication itself as a practice. For them, the success or failure of scientific domestication is framed as dependent not on human agency or evolutionary forces but on the invisible hand of the marketplace and on the inductive emergence of community values from individual reproductive decisions. This move can only be viewed in its historical context: global biotechnological markets are themselves currently reshaping critical social relationships— like those between humans and animals and between the public and private spheres. As Franklin observes: "We are currently witnessing the emergence of a new genomic governmentality. ... This is necessitated by the removal of genomes of plants, animals, and humans from the template of natural history that once secured their borders, and their reanimation as forms of corporate capital" (Franklin 2000: 188).

Current efforts to clone pet animals reflect this same approach. In February 2002, Mark Westhusin's laboratory announced that it had cloned a domestic cat, a tabby called "Cc" (short for "Copy Cat"). This achievement began as the "Missiplicity Project," a $3.7 million partnership between Texas A&M University's College of Veterinary Medicine and University of Phoenix founder John Sperling, aimed at cloning Sperling's beloved Siberian husky mutt, Missy. But it ended, in 2002, when Sperling dissolved this arrangement with the university, and reinvested in a private California-based company called "Genetic Savings and Clone" (or GSC). Westhusin suggests that medical benefits for humans might come from continued research on animal clones: "Cats have feline AIDS and that's a good model for studying human AIDS." Sperling, in turn, wants to put the ability to replace lost pets within the reach of all consumers. For an initial fee of about $1,000, plus a $100 annual maintenance fee, GSC customers can now have their cat or dog's DNA stored in the company's gene bank in anticipation of the day they can bring home a pet clone (target price tag: $20,000; "First Pet Clone is a Cat" 2002; Pray 2002) Either project, it seems, presumes a "maximal utility" relationship between humans and animals, in which (even though the motive is meant to confirm a caring ethos between human and animal companions) human money has all the power in shaping the outcome (McHugh 2002).

Likewise, work on "designer" mutant mice points to similar conclusions. In mass media representations such creatures sometimes appear as generic aggregates—"genes-for-X mice," in which for X, one can substitute medical conditions as diverse as Alzheimer's disease, male patterned baldness, obesity, and aggression (to name only a few). But occasionally such mice emerge as individual fait d'acompli, with unique names and commercial applications attached to their identities. One of the most widely reported of the latter was called "Doogie, the Smart Mouse." In mid-September 1999, Princeton University neurobiologist Joe Tsien created this animal, and named him after the TV child prodigy "Doogie Houser, MD." Doogie is a so-called "smart mouse" because his memory has been enhanced through genetic manipulation of NR2B, a gene whose product pairs with another gene's product, NR1, to open what biologists believe to be the physical mechanism of memory in the brain. Doogie's forebrain now produces some extra NR2B product, so his memory mechanism stays open an extra 1/150,000ths of a second (Wade 1999). This is enough time, Tsien and his colleagues say, for the creature to outperform other normal mice its age on "standard" tests of rodent intelligence. *Time Magazine* proclaimed that Tsien's work "sheds lights on how memory works and raises questions about whether we should use genetics to make people brainier" (Lemonick 1999: 54). Meanwhile, Princeton filed for a use patent on the NR2B gene, which would give the institution the right to develop drugs to enhance NR2B production in humans (Arnst 1999; cf. Kevles 2002; Zitner 2002). As a material incarnation of the practical and ideological values of genetic-based experimental biomedicine, then, designer mutants like Doogie are true domesticates. Doogie is part nonhuman animal, in so far as scientists have altered his genes and experimented on his body in ways ethically unthinkable for members of our own species but bearing a human name and performing functions significant only within our culture.

Haraway and Rabinow would argue this development signals that the terms of debates over controlled breeding are again themselves changing, because (as Donna Haraway argues) GE animals have, since WWII, inhabited more and more locations "where the categories of nature and culture implode" (1997: 255). Throughout the 20th century the lay public for the most part tolerated the ambiguity of scientific animal and scientific human coexistence. But because I am less concerned about the historical matter of "when" than in the "why" and "how" of this process, the ultimate questions seem to be these: As we broach new forms of genetic domestication, is the naturalization of the political and economic aspects of animal domestication in the laboratory that we have witnessed inevitable? What does this construction of the relationship between "science" and "domestication" make visible and what does it obscure?

Incongruous images and beliefs about the GE laboratory animals remain powerful as much for their ability to bring us together under an umbrella of common humanity they serve, as to tear us apart, in periodic waves of social and ethical conflict over their larger meanings. For earlier generations this tension was captured in Aldous Huxley's *Brave New World* (1932), a fictional human genetic dystopia founded on eugenic methods formerly believed appropriate only for farm animals, as well as (in the United States) the reality of "Model Sterlization Laws" enacted in 1914 (Huxley 1932; Lombardo 1988, 2003). Likewise, for our own era, ambivalence about the converging roles of human and nonanimals in the historical process of laboratory domestication is both fact and fiction: it is the finished code of the Human or Mouse or Dog Genome Project but also the technoscientific urban imaginary of Zadie Smith's novel *White Teeth* (2000), the climactic scene of which features the public launch of celebrated (fictional) scientist Marcus Chalfen's FutureMouse©.

Smith's FutureMouse© is a transgenic mouse in which fatal genes have been engineered to be "turned on" and expressed along a predictable timetable. This animal, Dr. Chalfen claims, is the "site for an experiment" into the aging of cells, the ultimate embodiment of human agency in the process of scientific development: "FutureMouse©, he tells the crowd, holds out the tantalizing promise of a new phase in human history where we are not the victims of the random but instead directors and arbitrators of our own fate." For Chalfen's middle- and working-class teenage children and their friends, however, FutureMouse© is a metaphor for the degree to which their father's personal affections for them (and by extension, his concern for the problems affecting contemporary Western, esp. British, society) have been distracted by science, technology, and other bourgeois values. Still, when asked by a group of animal activists (F.A.T.E.: for "Fighting Animal Torture and Exploitation") to liberate the animal, Chalfen's oldest son Josh (mothered by his horticulturalist wife Joyce and with whom Marcus has an especially tense relationship) surprises everyone by resisting his friends, and expressing what he holds to be a certain fatalism about this mouse's development: "this isn't like the other animals you bust out. It won't make any difference. The damage is done" (Smith 2000: 401, 357, respectively).

FutureMouse© does eventually escape, and the creature gladly seizes its elusive and short-lived freedom and (literally) runs with it. Still, in this one act of resistance, son Josh redeems his personal relationship with his father and seals the fate of FutureMouse© in the world—and in so doing, this fictional character encourages us to redirect our historical focus to what's really interesting and important about animal genetic engineering. Just as in *White Teeth* Smith is not as interested in the issue of race per se as she is the juxtaposition and interaction of urban people from different ethnic groups living their daily lives, Josh is not as interested in the GE animal per se as much as the juxtaposition and interaction of animals and people domesticated by science as a metaphor for life as it is, rather than life as we wish it to be.

To the extent that history informs, and is itself informed by, the present, these various scientific domestication projects—past and present, fictional and actual—demonstrate that all animals and humans are contested actors in a contingent process. Remarking on the genetic similarities between mice and humans, yeast geneticist Ira Herkowitz recently said: "I don't consider the [laboratory] mouse a model organism. The mouse is just a cuter version of a human, a pocket-sized human" (Hall n.d.). If we are to formulate a serious conversation about domestication and its implications for the relations of postmodern biopower, then we must acknowledge that such translations of the domestication metaphor cut both ways. If (as Jean Baudrillard infamously declared) "Disneyland exists to conceal the fact that America is Disneyland" then the laboratory mouse exists to conceal the fact that man is also a mouse (cf. Holthof 1993)—and Herkowitz's easy scientific conversion makes this translation visible. But likewise: although neither is the "postgenomic" era as new as it is made to seem nor does the continued use of the metaphor of scientific domestication condemn us to merely fulfill a prophecy of biological determinism or history's Progressive-era vision of eugenic governmentality. Recognizing the prominence of the domestication metaphor in genetics, past and present, calls us to begin the difficult, but necessary, task of accounting for and intervening in the complicated process of technical, cultural, and political formation that its ongoing uses shape for humans and animals. In so doing, we reanimate our understanding of domestication in both anthropology and in history— and create new opportunities for the important policy conversations that (in often-polarized debates over animal use in science) have thus far proven elusive.

Notes

1. "That's a bit of an overstatement. We've been in the pre-genome era for all of human history. On April 14, we will finally have our genome. ... We ought to put the nix on post-genomics. Otherwise, we may fool ourselves that we know more about our genome than we do" (Collins 2003).

2. A search of the *Reader's Guide to Periodical Literature* for the years 1927 to 1939 turn up the following accounts to illustrate the point: "House Mice Indicated as Nerve Disease Carriers" 1939, "Rodent Control in Food Establishments" 1937, Johnson 1929, "Mice as Disease Carriers" 1929, "Are Mice Smarter Than Men?" 1928, and Humphrey 1927. For a more exhaustive analysis of a similar cultural phenomenon with rats, see Hendrickson 1988.

3. Interestingly, Disney's original choice for a new character that year was not a mouse but a rabbit. But when contract disputes forced him to quit his original company, which owned all of his extant work, he needed to develop a new figure and he supposedly drew Mickey from his own experiences with mice: "He liked mice; in fact, he had sheltered some wild mice in his office in Kansas" (Larson 1974: 18; see also Bain and Harris 1977: 10–16). But see "Mickey Mouse's Miraculous Movie Monkey-Shines" 1930, "Mickey Mouse is Eight Years Old" 1936.

4. Lathrop's contributions to the historical development of the inbred mouse are acknowledged in the second edition of *The Biology of the Laboratory Mouse* (Staff of the Jackson Laboratory 1966), in Potter and Lieberman 1967, in Morse 1978, and Klein 1975.

5. Any estimate of mouse use will be problematic: Rowan's sources, for example, are unclear and he does not break down the kinds of mice included in that percentage (namely, percentage of inbreds vs. outbreds). But the best statistics on animal use at the U.S. NIH confirm his conclusions: mice and rats together, for example, account for 60–70 percent of all animals used, and approximately 90 percent of all mammals used. An additional problem is one of scientific animal accounting: who, for example, is compiling numbers for every use of a *Drosophila* fly in a test tube? The existence of such recordings would drastically change animal use estimates.

References

"Are mice smarter than men?" 1928. *Popular Mechanics* 49: 519.

Arnst, Catherine. 1999. Building a smarter mouse. *Business Week* 3646 (September 13): 103.

Arthus-Bertrand, Yann. 1999. *Good breeding.* New York: Harry N. Abrams.

Bain, David, and Bruce Harris. 1977. *Mickey Mouse: Fifty happy years.* New York: Harmony Books.

Becker, Alida. 2000. Man and beast. *New York Times Book Review,* Late Edition, March 19, Section 7, Column 2: 19.

Beckman, Bjarne. 1974. *Von Mausen und Menschen: Die hoch- und spatmittelalterlichen Mausesagen mit Kommentar und Anmerkungen.* Bremgarten: Beckman.

Best in Show. 2002. Christopher Guest, dir. 90 min. Warner Brothers. Burbank.

Birkhead, Tim. 2003. *A brand new bird: How two amateur scientists created the first genetically engineered animal.* New York: Basic Books.

Budiansky, Stephen. 1992. *The covenant of the wild: Why animals chose domestication.* New Haven, CT: Yale University Press.

Collins, Francis. 2003. Keynote Speech, April 13 Boston Bio-IT World Expo. (author's notes)

de Gubermatis, Angelo. 1968 [1872]. *Zoological mythology, or the legends of the animals.* Detroit: Singing Tree Press.

Holt, Sarah, and P. Williams, prods. 2003. *Eighteen ways to make a baby. NOVA* special, PBS, June 10th.

"The English craze for mice." 1937. *Reader's Digest* 30 (March): 19.

"First pet clone is a cat." 2002. *BBC News Online,* February 15. Available at: http://news.bbc.co.uk/1/hi/sci/tech/1820749.stm, accessed September 9, 2003.

Franklin, Sarah. 1997. *Embodied progress: A cultural account of assisted reproduction.* New York: Routledge.

——. 2000. Life itself: Global nature and the genetics imaginary. In *Global nature, Global culture,* edited by Sarah Franklin, Celia Lury, and Jackie Stacey, 188–227. London: Sage.

——. 2003. Ethical biocapital: New strategies of cell culture. In *Remaking life and death: Towards an anthropology of the biosciences,* edited by Sarah Franklin and Margaret Lock, 97–128. Santa Fe: School of American Research Press.

Galton, Francis. 1865a. The first steps towards the domestication of animals. *Transactions of the Ethological Society of London* 3: 122–138.

——. 1865b. Hereditary talent and character. *Macmillan's Magazine* 12: 157–166, 318–327.

——. 1883. *Inquiries into human faculty and its development.* London: J. M. Dent and Sons.

Ginsburg, Faye, and Rayna Rapp, eds. 1995. *Conceiving the new world order: The global politics of reproduction.* Berkeley: University of California Press.

Grohmann, Josef V. 1862. *Apollo Smintheus und die bedeutung der mause in der mythologie der Indogermanen*. Prague: J. C. Calve.

Grüneberg, Hans. 1943. *The genetics of the mouse*. Cambridge: Cambridge University Press.

Hall, Stephen S. n.d. *Our closest relative among model organisms. The mouse sequence: A Rosetta Stone*. Howard Hughes Medical Institute. Available at http://www.hhmi.org/genesweshare/d100.html, accessed September 22, 2006.

Haraway, Donna. 1997. *Modest_witness@second_millennium.FemaleMan© _meets_OncoMouse™: Feminism and technoscience*. New York: Routledge.

——. 2003. Cloning Mutts, Saving Tigers. In *Remaking life and death: Towards an anthropology of the biosciences*, edited by Sarah Franklin and Margaret Lock, 293–328. Santa Fe: School of American Research Press.

Hendrickson, Robert. 1988. *More cunning than man: A social history of rats and men*. New York: Dorset Press.

Holthof, Marc. 1993. To the realm of fables: The animal fables from Mesopotamia to Disneyland (translated from the Dutch by Milt Papatheophanes). In *Zoology on (Post) Modern Animals*, edited by Bart Verschaffel and Marc Vermick, 37–55. Dublin: Lilliput Press.

"House mice indicated as nerve disease carriers." 1939. *Science News Letter* 35: 329.

Humphrey, F. T. 1927. Serious plague of mice in California. *Scientific American* 136: 330–331.

Huxley, Aldous. 1932. *Brave new world, A novel*. New York: Harper and Row.

Johnson, Rhea Kimberly. 1929. White mice. *Nature Magazine* 13: 48–50.

Keeler, Clyde. 1931. *The laboratory mouse: Its origin, heredity and culture*. Cambridge, MA: Harvard University Press.

——. 1932. In quest of Apollo's sacred white mice. *Scientific Monthly* 34 (January): 48–53.

Kevles, Daniel J. 2002. "Of mice and money: The story of the world's first animal patent." *Daedalus* 131 (2 [Spring]): 78–88.

Kimmelman, Barbara. 2003. *Mendel finds a home: Disciplinary momentum in the American reception of Mendel, 1900–1910*. Paper presented at the International Society of History, Philosophy, and Social Study of Biology Meeting, in Vienna, Austria, July 16–20.

Klein, Jan. 1975. *The biology of the histocompatibility-2 complex of the mouse*. New York: Springer-Verlag.

Kohler, Robert. 1995. *Lords of the fly*. Chicago: University of Chicago Press.

Larson, Norita. 1974. *Walt Disney: An American original*. Mankato, MN: Creative Education.

Leach, Helen M. K. 2003. Human domestication reconsidered. *Current Anthropology* 44 (3): 349–368.

Lemonick, Michael D. 1999. Smart genes? *Time Magazine* 154 (September 13): 54–59.

Little, Clarence Cook. 1916. The relation of heredity to cancer in man and animals. *Scientific Monthly* 3: 196–202.

——. 1935. A new deal for mice: Why mice are used in research on human diseases. *Scientific American* 152: 16–18.

Lombardo, Paul. 1988. Miscegenation, eugenics, and racism: Historical footnotes to Loving v. Virginia. *University of California Davis Law Review* 21: 421–452.

——. 2003. "Eugenic sterilization laws." Essay for the Human Genome Project eugenics archive. Available at: http://www.eugenicsarchive. org/html/eugenics/essay8text.html, accessed December 19.

Malriey, Pierre. 1987. *Le bestiare insolite: L'animal dans la tradition, le mythe, le rêve*. Realmont, France: Editions la Duralie.

McHugh, Susan. 2002. Bitches from Brazil: Cloning and owning dogs through the Missyplicity Project. In *Representing animals*, edited by Nigel Rothfels, 180–98. Bloomington: Indiana University Press.

"Mice as disease carriers." 1929. *Science* 69 (Supp. 14): 1631.

"Mice beautiful." 1937. *Time Magazine* 30 (July 19): 50.

"Mickey Mouse is eight years old." 1936. *Literary Digest* 122: 18–19.

"Mickey Mouse's miraculous movie monkey-shines." 1930. *The Literary Digest* 6: 36–37.

Midgley, Mary. 1983. *Animals and why they matter*. Athens: University of Georgia Press.

Morse, Herbert S., ed. 1978. *The origins of inbred mice: Proceedings of an NIH workshop, Bethesda, Maryland*. New York: Academic Press.

"Mouse Show." 1937. *Newsweek* 9 (January 23): 40.

Moulton, Hope. 1901. Pestilence and mice. *Classical Review* 15: 284.

Muller, H. J. 1928. The production of mutations by X-rays. *Proceedings of the National Academy of Sciences* 14: 714–726.

National Institutes of Health (NIH). 1990. *Office of Technology assessment: Report on lab animal use*. Bethesda, MD: NIH. (provided to author in 1993 by Dr. Louis Siebel)

New York Times. 1996. New breed of mouse aids Alzheimer's work. *New York Times*, October 4: A22.

Paul, Diane, and Barbara Kimmelman. 1988. Mendel in America: Theory and practice, 1900–1919. In *The American development of biology*, edited by R. Rainger, K. Benson, and J. Maienschein, 281–310. New Brunswick, NJ: Rutgers University Press.

Pauly, Philip J. 2000. *Biologists and the promise of American life: From Meriwether Lewis to Alfred Kinsey.* Princeton: Princeton University Press.

Pew Initiative on Food and Biotechnology. 2002a. *Animal cloning and the production of food products—Perspectives from the food chain* (September 26). Available at: http://pewagbiotech.org/events/0924, accessed June 1, 2004.

———. 2002b. *Biotech in the barnyard: Implications of genetically engineered animals (September 24–25).* Available at: http://pewagbiotech.org/events/0924, accessed June 1, 2004.

Pincus, Gregory, and B. Saunders. 1939. The comparative behavior of mammalian eggs in vivo and in vitro. *Anatomical Record* 75: 537–545.

Potter, M., and R. Lieberman. 1967. Genetics of immunoglobulins in the mouse. *Advances in Immunology* 7: 91.

Pray, Leslie. 2002. Missyplicity goes commercial. *The Scientist,* November 27. Available at: http://www.the-scientist.com/news/20021127/03, accessed December 5, 2005.

Provine, William B. 1989. *Sewall Wright and evolutionary biology.* Chicago: University of Chicago Press.

Rabinow, Paul. 2003. *Anthropos today: Reflections on modern equipment.* Princeton: Princeton University Press.

Rader, Karen A. 1999. Of mice, medicine, and genetics: C. C. Little's creation of the inbred laboratory mouse, 1909–1918. *Studies in the History and Philosophy of Biology and the Biomedical Sciences* 30 (3): 319–343.

———. 2004. *Making mice: Standardizing animals for American biomedical research.* Princeton: Princeton University Press.

"Rodent Control in Food Establishments." 1937. *American Journal of Public Health* 27: 62–66.

Rowan, Andrew. 1984. *Of mice, models and men.* New York: State University of New York Press.

Russell, Nerissa. 2002. The wild side of domestication. *Society and Animals* 10 (3): 285–302.

Schloegel, Judith Johns. 2006. *Intimate biology: Herbert Spencer Jennings, Tracy Sonneborn, and the career of American protozoan genetics.* Ph.D. dissertation, Department of History and Philosophy of Science, Indiana University.

Silver, Lee. 2001. Thinking twice, or thrice, about cloning. In *The cloning sourcebook,* edited by Arlene J. Klotzko, 60–63. New York: Oxford University Press.

Smith, Zadie. 2000. *White teeth: A novel.* New York: Random House.

Springfield Sunday Republican. 1913. [No title—feature article about Abbie Lathrop]. *Springfield Sunday Republican,* October 5: 12.

Strathern, Marilyn. 1992. *Reproducing the future: Essays on anthropology, kinship, and the new reproductive technologies.* New York: Routledge.

Staff of the Jackson Laboratory. 1966. *The biology of the laboratory mouse.* New York: McGraw Hill.

Stepan, Nancy Leys. 1986. Race and gender: The role of analogy in science. *Isis* 77 (2): 261–277.

Thompson, Charis Cussins. 2001. Strategic naturalizing: Kinship in an infertility clinic. In *Relative values: Reconfiguring kinship studies,* edited by Sarah Franklin and Susan McKinnon, 175–202. Durham, NC: Duke University Press.

Trudinger, Melissa. 2003. Domestication: It's only natural. *Australian Biotechnology News,* July 7. Available at: http://www.biotechnews.com.au/index.php?id=135033282, accessed December 5, 2005.

Wade, Nicholas. 1999. Of smart mice and an even smarter man. *New York Times,* 7 September, Late Edition—Final: F1.

Wilmut, Ian, Keith Campbell, and Colin Tudge. 2000. *Second creation: Dolly and the age of biological control.* New York: Farrar, Strauss, and Giroux.

Zitner, Aaron. 2002. Patently provoking a debate. *Los Angeles Times* May 12: A1+.

Domestication "Downunder": Atlantic Salmon Farming in Tasmania

Marianne Lien

Could intensive aquaculture be seen as a recent kind of domestication? To what extent can farmed salmon be compared with chicken or beef? This chapter seeks to explore the ways in which marine farming differs from its terrestrial counterpart, arguing that such a comparison may shed light on contemporary aquaculture as well as challenging the cultural distinction between the domestic and the wild that underpins the notion of domestication.

Marine Domestication: A Historical–Comparative Approach

Aquaculture is an ancient practice, and the earliest references to pond-fish culture date back several millennia (FAO 2003). Yet, until the 1970s, the overall contribution of aquaculture to the human diet was limited. Within the last three decades, however, world aquaculture production has increased exponentially, contributing to what some refer to as the "blue revolution." Aquaculture is now among the fastest growing food production sectors in many countries in the world, and almost a third of all the fish that is consumed by humans today has been raised on a fish farm (FAO 2003). This dramatic increase in aquaculture production during the last decades is closely connected to an increase in so-called intensive systems of fish farming, which are the focus of this chapter.

The emergence of intensive fish farming implies that a range of marine species that have, until recently, been living "in the wild" are now domesticated for commercial purposes and raised for human consumption on a massive scale. Although the future prospects of aquaculture are necessarily uncertain, I suggest that what we witness today as intensive aquaculture production represents a significant recent turn in the human history of domestication. What started with the cultivation of plants in the Middle East 11000 B.C.E. and was followed by the domestication of terrestrial species around 9000 B.C.E. (Leach 2003),[1] is now extended to the domestication of aquatic species on a massive scale.

All modes of subsistence may be conceptualized as specific modes of production, situated in historical and evolutionary context (Ellen 1994: 198). Aquaculture is often classified as a subcategory of fishery production, and is, by default, juxtaposed or compared to the more familiar and traditional domain of (wild) capture fisheries. However, as soon as we focus on aquaculture from the perspective of domestication, other continuities emerge, and aquaculture appears as a kind of husbandry, comparable to terrestrial agriculture.

In this chapter, I approach aquaculture as a kind of domestication "under water," an approach that invites a range of questions pertaining to both contemporary practices and historical development. We may explore, for instance, the implications of a shift from a terrestrial to an aquatic environment focusing on the socially constituted structures within which domestication takes place and ask: What are the differences between domestication in soil or air and in water in terms of knowledge, control and human–animal interaction?[2] What are the limits in relation to the transfer of technological, organizational, and judicial regimes of production from terrestrial agriculture to aquaculture? How are human–animal relations constituted and established across the boundary of the water surface?

Approaching aquaculture as a recent turn in the history of domestication also invites comparison along a historical dimension. Although knowledge of the agricultural revolution is only available to us through historical and archeological sources, we may approach the recent emergence of intensive aquaculture as a key event in the human history of domestication, which we, just now, are uniquely positioned to study. Thus, we may approach aquaculture as "historians of the present," asking: To what extent is intensive aquaculture modeled on contemporary animal husbandry or modern agriculture? In what ways are contemporary features related to terrestrial agriculture "built into" the technoscientific regimes of intensive aquaculture? How do contemporary notions of nature, territorial bioboundaries, and biotic invasion structure or inform aquaculture practices? In other words, how may the current turn toward marine domestication be analyzed as a product of the specific historical circumstances during which this shift takes place? Although aquaculture clearly has a place within the field of maritime anthropology (cf. Ballinger and Helmreich n.d.), the questions above indicate that it also challenges the boundaries that underpin such a subdiscipline. In this chapter, I therefore draw on studies of human–animal relations in general, both from the marine and from the terrestrial domain.

Local Variation and Global Networks

Just as the historical pathways to agriculture are several and vary significantly (Russell 2002), the domestication of aquatic species also takes a number of different forms. As such, it is a mode of subsistence that cannot be understood:

> except as part of ... a socially constituted structure, nor can it be approached analytically apart from this context, for it is inevitably a consequence of social action which is in part purposive, and has its origins in particular social relations of appropriation. (Ellen 1994: 198)

Different definitions of domestication highlight different aspects of such relations of appropriation (see Russell this volume). Approaching aquaculture as domestication, I rely on a much-cited definition, emphasizing the economic aspects of the relation, according to which a domesticated animal is such:

> one that has been bred in captivity, for purposes of subsistence or profit, in a human community that maintains complete mastery over its breeding, organization of territory and food supply. (Clutton-Brock 1994: 26)

This approach implies that other dimensions frequently involved in processes of domestication, such as taming and the sharing of *domus* (see Wilson this volume) are not emphasized. These dimensions are not easily accommodated in intensive aquaculture (see discussion below).

The expansion of intensive aquaculture has taken place during a historical phase characterized by extensive global networks, and the recent expansion of aquaculture is both dependent on and contributing to transnational connections through the transnational transfer of technology, know-how, capital, feed, and even genetic input. Yet every marine operation is necessarily a localized activity. Although a salmon farm on the Norwegian coast may appear at first sight to be almost identical to a similar-sized farm in Canada, Chile, or Tasmania, its location defines, to some extent, its enrolment in ongoing alliances, conflicts, and aims. Furthermore, it is subject to a marine environment of a particular kind, to legal regulations and socioeconomic patterns of production that differ nationally and/or regionally, and to interpretations of landscapes and seascapes that are local yet often sustained by transnational notions of nature, biodiversity, and environmentalism. Intensive aquaculture bears the mark of an historical era of enhanced transnational flows (of people, ideas, things), and one in which cultural communities are not always bounded by territorial space.

The recent expansion of intensive aquaculture in the form of Atlantic salmon farming may be seen as part of a more general global expansion of markets in the coastal domain, and of the commoditization of marine resources because of the social and economic reorganization of capture fisheries (Pálsson 1991). It also coincides with a period of overexploitation of global fish stocks, and an increasing concern about the depletion of global marine resources (Apostle, McKay, and Mikalsen 2002). The emergence of intensive Atlantic salmon farming was not directly a result of the decline of the fisheries (unlike on the Pacific coast of North America, salmon was abundant in Norwegian rivers in the 1970s, and still is). Nevertheless, in contemporary public and scientific debates today, aquaculture is often portrayed both as the cause of, and the solution to, such decline. The former argument relates primarily to the farming of carnivorous marine species such as salmon, and their requirements of feed based on fish oil, which contributes to an increased demand on the global stock of small marine species that feed low in the food chain (anchoveta, herring, capelin, etc.; Naylor et al. 2000). As the availability of such pelagic fish stocks is limited (many are already classified as overexploited), some of the fish oil is now replaced with plant-based oils in salmon feed, and research is directed toward further development of alternative feed sources. The latter argument is based on what is seen as the untapped potential of aquaculture in providing fish for human consumption and in feeding a growing human population, and often involves the argument that farmed fish are more efficient converters of feed than fish in the wild. The impact of salmon farming on world fish supplies is, however, subject to considerable debate, and is beyond the scope of this chapter (for an overview see Naylor et al. 2000).

In this chapter, I focus on aquaculture as it unfolds today in Tasmania, Australia, while the history of aquaculture in Norway serves as a comparative backdrop for the analysis. The ethnographic material from Tasmania is based on fieldwork in 2002 and 2004, while more general claims about aquaculture and salmon farming are based on the study of secondary sources available through libraries, trade magazines, and the Internet. As the title indicates, an exploration of domestication "downunder" involves a dual ambition. Although I seek to explore the general characteristics of husbandry *under water*, I simultaneously engage a *particular locality*, that is an Australian island state located at 40–42° S in the Southern Sea. As I show, this choice of location offers particular challenges in relation to the classification of domesticated salmon in relation to coastal landscapes and seascapes.

The discussion that follows is divided in three parts. First, I discuss the conceptualization of domesticated salmon in relation to existing dichotomies of the wild and domestic and between native and introduced as they operate in Norway and in Tasmania respectively. Second, I discuss the implications of the water surface as a boundary between humans and domesticated salmon, and the strategies that are used to establish a technoscientific interface. Third, I draw attention to the fluid quality of water, and discuss how salmon farming is characterized by confinement in particular forms of marine space, that have a physical, a legal, and a symbolic dimension. But first, I give a brief account of the domestication of Atlantic salmon in a historical perspective, and the more recent domestication of salmon on the Tasmanian coast.

Domestication in its Infancy: The Story of Atlantic Salmon

Salmon have a longer history on earth than human beings. They belong to a group of fish known as the Teleostei that first appeared 150 million years ago, during the Jurassic, and prior to the emergence of mammals. Teleost fish now occupy all aquatic environments on earth, and show a tremendous variation. Amongst these, salmon has a very basic and primitive body form that appears to have changed little since the Jurassic (Stead and Laird 2002: 1–2).

Whereas most terrestrial farm animals have been selectively bred and farmed since Neolithic times, the cultivation of the entire life cycle of Atlantic salmon *(Salmo salar)* in captivity has only been widely practiced for the last three decades (Stead and Laird 2002). This difference is significant, because it implies that although the former share a history of coevolution with humans and have adapted genetically to life in captivity for hundreds or even thousands of years, salmon are still "newcomers to the farm." Furthermore, although pond aquaculture usually involves species adapted to freshwater (carp, trout), Atlantic salmon are anadromous, that is, they are genetically predisposed to a life that involves long-distance migration from rivers to the ocean and back. Unlike most other forms of aquaculture, the commercial domestication of Atlantic salmon involves two different phases: hatching and early growth in freshwater environment, and subsequent growth in marine farms on the ocean. This implies phases of transfer and adaptation during which salmon are particularly vulnerable, and with which humans have relatively little experience. In other words, salmon farming represents marine domestication in its early infancy.

The first successful attempts to domesticate salmon took place in Norway in the late 1960s. The industry grew rapidly, and by the 1990s, Norway had become a world leading producer and exporter of Atlantic salmon (Kaiser 1997). Since then, the global production of Atlantic salmon has exceeded all expectations. Today, production of Atlantic salmon has moved far beyond the species native habitat, and Chile is now among the major suppliers in the world. Other producers include Canada, Ireland, Great Britain, and Australia. Farmed Atlantic salmon is the most important product within the category of salmonids, accounting for just above 1 million tons in 2001 (FAO Yearbook 2001). As the population of wild salmon tend to decline, a rough estimate indicates that more than 94 percent of all the world's adult Atlantic salmon occupy the aquaculture niche (Gross 1998).

Contemporary salmon farming is an intensive form of aquaculture that implies that successive generations of fish are raised in captivity during their entire lifecycle under a regime of strict control of feeding and reproduction. It differs from pond aquaculture in that the production is highly commercial and more capital intensive. It also differs from the less intensive human interventions in rivers and streams to enhance the fish stock for sports fishing in that intensive marine farming allows no planned release of fish from captivity to the wild, and that all phases of the farmed salmon lifecycle are monitored and facilitated by human management. Thus, the physical boundary between "domestic" and "wild" is systematically enforced. This is reflected in the use of language of salmon farming. Aquaculture grow-out units are usually called "farms," and live farmed salmon are ultimately "slaughtered" or "harvested," terms that draw on the agricultural domain. Wild salmon, on the contrary, are "caught," a term associated with hunting and fishing (see also Pálsson 1991).

The difference between river management and intensive fish farming reflects the distinction between "protection" and "domestication" as two different categories of human–animal relations. According to Harris (1996), *protection* implies human intervention that seeks to modify the predator–prey relationship in favor of a particular species by enhancing the reproductive potential of the species through some manipulation of its environment, or by taking a nucleus of animals into captivity to reproduce. This typically describes the management of salmon rivers in North America, Britain, and Norway in the 19th and 20th century (stocking of rivers, building of salmon passes), and indicates that what is marketed as "wild salmon" is a species that has been shaped by human interventions for at least 150 years (cf. Montgomery 2003). Domestication, on the contrary, is defined as such:

> the maintenance by humans of a self-perpetuating breeding population of animals isolated genetically from their wild relatives, with resulting behavioral, and usually also phenotypic, changes in the domestic stock. (Harris 1996: 454)

This definition highlights how, during the course of domestication, selective mechanisms have led to genetic changes that make the domesticated species dependent on humans for their survival. Recent studies indicate that similar adaptive processes take place for cultivated fish, as the differences in the environment experienced by wild and cultured fishes offer considerable scope for unplanned natural selection, as well as behavioral differences arising from different experiences (Huntingford 2004). In the case of Atlantic salmon *(S. salar),* a wide range of genetic and developmental characteristics now differentiate the farmed salmon from the wild, including more aggressive behavior, increased stress response and a poorly developed antipredator response (Gross 1998; Huntingford 2004). This has led some authors to propose the term *S. domesticus* to indicate the emergence of a distinct evolutionary lineage, and to help us realize that in spite of the large biomass of *Salmo* in aquaculture, the wild species of *S. salar* is endangered (Gross 1998: 133).

It has been argued that a complete process of domestication involves the extermination of the wild ancestors of the domesticated species (Zeuner 1963), as is more or less the case for poultry, sheep, and cattle in Europe. In the case of salmon, its short history of domestication implies that these criteria are not fully met. The above mentioned differences notwithstanding, farmed salmon still resemble salmon in the wild, they may escape and survive without human protection, and they are capable of breeding with individuals in the wild salmon in North Atlantic rivers (Utter and Epifanio 2002). Thus, the very recency of intensive marine (salmon) farming makes the distinction between domestic and wild fish more unstable than in the case of the more common farm animals, as farmed salmon transgress both spatial and conceptual boundaries between the wild and the domestic.

A great part of all Atlantic salmon aquaculture takes place along the North Atlantic rim, which is also the native habitat of wild salmon. In these regions, one of the key environmental concerns regarding salmon farming is the problem of salmon escaping from cages, and the concomitant uncertainty regarding the escapees' survival and mating behavior in rivers and streams, which represents a threat to the genetic diversity of the wild salmon populations (Kaiser 1997).

In light of the unique biodiversity of wild salmon in Norwegian rivers, it might be argued that intensive salmon farming ought to be located anywhere *except* on the Norwegian coast where escapees are within easy reach of potential wild mates, and where infectious diseases that affect farmed salmon may harm the wild population. With this in mind, I assumed that Atlantic salmon farming in Tasmania would be a less complicated enterprise, less entangled in environmental concerns. However, as we shall see, this is not the case.

Atlantic Salmon on the Tasmanian Coast

Domestication of Atlantic salmon on the Tasmanian coast began in the mid-1980s, when local attempts to raise rainbow trout on marine farms had already proven successful, and some experience with oyster farming was established. Rural areas were marked by unemployment, and further growth of aquaculture was seen by local authorities as an alternative to agriculture in peripheral areas. In 1982, a Norwegian professor called Harald Skjervold, who had made a career of genetic improvement of Norwegian cattle through selective breeding and who had recently advised the Norwegian government on the development of salmon aquaculture, visited Australia to advise on cattle genetics. He was invited to Tasmania as well, and had a meeting with researchers at the Tasmanian Fisheries Research laboratories.

Today, this visit is often retold as part of the narrative of how Tasmanian salmon farming was initiated. Skjervold, who also had extensive experience with salmon farming in Norway, claimed that the Tasmanian coast was ideally suited to the culture of salmon. He facilitated contact with Norwegian investors, and within a couple of years, a commercial enterprise was set up as a joint venture between the Tasmanian government and "Noraqua," a company that was established for this purpose with Norwegian capital.[3] Noraqua facilitated the floating of "Tassal," a marine farming enterprise that has been a leading producer of Tasmanian salmonids ever since. During the next few years, the production of salmon increased, and in 1989, following a downturn on the Oslo property market, Norwegian interests sold out. Since then, Tasmanian marine aquaculture has been, for a large part, Australian owned. In 2001–02 the total production of Atlantic salmon in Tasmania was slightly less than 15 thousand tons, making Tasmania a very minor supplier globally, but the major supplier of salmon to the Australian market. Practically all Australian farmed salmon is raised in Tasmania.

To start commercial salmon farming in Tasmania, salmon broodstock was needed. In 1983, a strict regime of Australian quarantine prevented free import of live fish into Australia. Consequently, a suitable stock of salmon was sought within the Australian national territory. As it turned out, a strain of Atlantic salmon that had originally been translocated from Nova Scotia in 1964 was found in New South Wales where it had reproduced as a landlocked population. Because of strict import regulations, this genetic stock has become the backbone of Tasmanian Atlantic salmon, and has remained practically unchanged ever since. Incidentally, having been taken to a place where it did not exist in the first place, Tasmanian farmed salmon may be seen as more truly domesticated than farmed salmon in Norway and Scotland. This is because domestication is fundamentally tied to *relocation*.

Between Wild and Domestic—The Cultural Classification of Atlantic Salmon

Western perceptions of nature have evolved through repeated juxtapositions in which nature is increasingly seen as something pure, untouched by human activity (Macnaghten and Urry 1998). One such juxtaposition is the conceptual distinction between "wild" and "domestic," according to which nature is associated with the wild, whereas domestication involves human society.

In Northern Europe, salmon has traditionally occupied the wild end of this continuum. In spite of the human effort that has traditionally been spent in protecting salmon stock (see above), salmon has been associated with nature, sportsmanship, and the outdoors, making salmon what some have called the "King of Fish" (Lien 2005).[4] Farmed salmon disturbs this image, and thus also the conceptual distinction between the domestic and the wild. Not only does it imply the domestication of a species metaphorically associated with the wild but it does so in a way that makes salmon comparable to industrially raised pigs or battery hens, animals that often serve to exemplify the ill effects of large-scale industrial farming. Yet, although terrestrial species such as goats, cattle, and pigs have evolved significantly as a result of domestication (Leach 2003), and parted genetically with their nondomesticated relatives some 7,000–9,000 years ago (Reitz and Wing 1999), farmed salmon are still so similar to the wild that it takes a trained eye to tell the difference. Unlike most farmed animals, salmon that escape survive without human protection, and are, as mentioned, found to migrate to rivers where they may reproduce and thus contribute to a hybrid stock that is neither domestic nor absolutely wild (Kaiser 1997). Farmed salmon may thus be seen as an example of an anomalous species, as it escapes (symbolically *and* literally) its culturally defined domain of the wild and threatens the classificatory order (Douglas 1975). Like the African pangolin, which is an anomaly among the Lele (an animal found in trees "with the body and tail of a fish, covered in scales," Douglas 1975: 33), farmed salmon is neither wild fish nor fully domesticated, but something in between. The ambiguous quality of farmed salmon must be seen in relation to its position as a "newcomer to the farm." Unlike sheep, pigs, and cattle, farmed salmon does not (yet) enjoy the cultural familiarity that long term coevolution brings about. The fact that farmed salmon may occasionally cross the physical and genetic boundaries of captivity further sustains a Western perception of farmed salmon as somewhat ambiguous. These transgressions do not, however, seem to challenge the notional opposition of wild and domesticated as an underlying order in the classification of nature on the North Atlantic rim.

However, although the opposition between wild and domestic serves as a key distinction in relation to salmon in Northern Europe, this is not the case in relation to Tasmania Atlantic salmon. First of all, there are no wild Atlantic salmon in the Southern seas, that is, no wild counterpart that could be genetically altered through crossbreeding. Furthermore, there are few indications that a feral population could ever be established (Lever 1996). In spite of numerous attempts to naturalize wild Atlantic salmon in Tasmania in the late 1800s, salmon turned out to be unable to reproduce (Lien 2005). Escapees may survive in the estuaries, but have not succeeded in returning to Tasmanian rivers to spawn. In other words, there is only one kind of salmon in Tasmania, and that is domesticated Atlantic salmon.

Second, although ideas of nature in Northern Europe are structured by the underlying dichotomy of wild–domesticated, Tasmanian ideas of nature are structured also in relation to a very different distinction, namely between the native and exotic (Low 1999, Trigger and Mulcock 2005). This implies that nature is ordered first and foremost along a dimension of genetic or geographic origin rather than along a continuum of domestication. Although the origin of a species is rarely at the forefront of environmental concerns in continental Northern Europe (perhaps because a gradual translocation of species has taken place in the region since the ice withdrew 10,000 years ago), knowledge of the origins of different species is central to environmental discourse in Tasmania, the rest of Australia, and other parts of the "neo-Europe" (Dunlap 1999). The story of the way European colonizers transformed the Australian landscapes is well-known, and has become, in many ways, a master narrative in Australian nature discourse (Clark 1999; Rolls 1969). As a consequence (sometimes unintended) of engagements with Australian soil, flora, and fauna, early colonizers contributed to disruptions of ecological connections that have, in turn, wiped out a great number of native species and thus contributed to a great loss of biodiversity. The late 18th century has thus become a historical "watershed" in distinguishing "native" Tasmanian nature from landscapes that bear the mark of human interference and biotic migration. This historical transformation, and the ways in which it is retold, inform contemporary orderings of nature, and sustain a distinction between native and "invasive" that orders practically all living species, with the possible exception of humans (cf. Head and Muir 2004). This distinction cuts across the juxtaposition between wild and domestic. Although the native is usually associated with the wild (wombat, wallaby, Tasmanian devil), and the introduced is often a domesticated animal (sheep, salmon, cattle), the native–invasive distinction also nurtures the possibility that wilderness may be transformed by domesticated invasive species that "go feral" and make the habitat of the natives uninhabitable (see also Woods and Moriarty 2001).[5] Because of these local entanglements, the domestication of Atlantic salmon in Tasmania is not as simple as it might seem from a Northern European perspective. Rather, Tasmanian Atlantic Salmon is caught up in local concerns that make domesticated salmon just as much a "matter out of place" in Tasmanian estuaries as in Norwegian fjords.

We have seen how some of the challenges involved in the domestication of salmon are caused by the way farmed salmon challenges the conceptual distinction between the domesticated and the wild in Western thought, and that this is because of, in part, the recency of the marine farming endeavor. This involves a specific valuation of landscapes and species that are seen as "wild," natural, or unaffected by human intervention. Such valuations are central to European ideas about nature (Harris 1996), and have inspired and informed environmental engagements far beyond its region of origin. Such concerns constitute part of the historical framework for the current expansion of intensive aquaculture.

With this cultural and contextual framework in mind, we may now turn to questions raised initially regarding the expansion from terrestrial to marine domestication, and explore in some detail the ways in which water "makes a difference." In the following, these differences are explored from two different angles. In the following section, I discuss the water surface as a boundary between humans and domesticated salmon, and as an interface in relation to human–animal interaction and control. Subsequently, I shall discuss the fluid qualities of coastal water in relation to environmental control and notions of property.

The Water Surface: Husbandry at a Distance

When Harald Skjervold suggested that the Tasmanian coast was ideal for salmon farming he was partly wrong. What might have appeared from above water as a topography resembling a Norwegian fjord was, in fact, a rather shallow estuary that lacked the depth and current to ensure the steady flow of water needed for intensive aquaculture. Today, the problem of shallow estuaries remains a major obstacle to the expansion of salmon farming along the Tasmanian coast. This brief narrative illustrates the extent to which experience and knowledge of salmon farming was, and still is, viewed as a transnational and general kind of knowledge that may be transferred from one geographical location to another (cf. Lien in press a). More importantly in light of the present discussion, it indicates how the physical conditions for marine farming are always slightly beyond the grasp of the human sensory perception.

Marine farming is an underwater activity, whereas humans are a terrestrial species. Intensive aquaculture, like intensive terrestrial farming, is fundamentally about *control* of the domesticated population. Consequently, to the extent that farmed salmon are mastered through regimes of control, they are necessarily mastered *at a distance*. In salmon farming, the issue of control involves practically all aspects of the lifecycle, but especially those that are related to growth and reproduction. In the following, I focus on feeding practices at the marine farm site, as this allows us to explore some of the ways in which "water makes a difference."

Contemporary salmon farming is a commercial enterprise firmly entrenched in production for profit through market exchange. Profit is a function of the amount of biomass produced when running costs are taken into account. The most important running cost of a marine farm operation is the cost of feed, usually accounting for more than 50 percent of total expenses. The challenge in salmon farming husbandry is, thus, to ensure that the salmon grow at an optimal rate with a minimum input of feed. Very crudely, the "task" given to each individual salmon is to grow (expand its biomass) as fast as possible, a process that depends on optimal feeding behavior and efficient metabolism. Yet, feeding and growth are complex processes, and a range of factors are likely to interfere. A successful regime of control thus involves a constant dialectic between monitoring the state of being of the population and administering an adequate response. Yet, this dialectic must somehow overcome the distance implied by the water surface. The ability of the farmhand to judge, for instance, the moment of satiety in the salmon pen, involves the interpretation of a number of different signs from beyond or across the water surface. Knowing when a group of salmon feeding perhaps four meters below the water surface has reached satiety represents a considerable challenge. This is partly because of the way in which the presence of a water surface between farm hand and fish cuts opportunities for sensing each others' movement and behavior (for a more detailed discussion, see Lien in press a).

In a chicken pen, as in most terrestrial contexts of domestication, each individual animal is readily exposed to the human caretaker as they are immediately perceived through the bodily senses (Ingold 2000). One may observe the way the chicken moves, hear the chicken cackle, notice their smell, and feel the softness of feathers or the heat produced by their bodies. Similarly, the chicken (or other domesticated species) may sense the presence of their caretaker and respond to him or her. Together, they produce a specific interactive pattern consisting of humans and animals. As human and terrestrial domesticated animals inhabit a space in which they are mutually exposed and to some extent intelligible to each other, such unmediated interaction is readily achieved. If we add to this the long history of coevolution of humans and certain domesticated animals, we may even speculate that perceptions and behaviors have been modified in both human and nonhuman domesticated species to accommodate a more nuanced *knowing* of the other (Leach 2003). As a result, a farm hand's decision to stop feeding a terrestrial farmed animal is facilitated by a rich array of signs made available to him through his or her bodily senses.

On a salmon farm, such opportunities for unmediated interaction are severely limited. Trapped by oxygen metabolisms, which are uniquely equipped for the elements of air and water respectively, human and fish are bound to occupy different physical spaces. Furthermore, as human sensory perception is severely diminished by the presence of a water surface (we do not see very far, hear very well, can rarely feel, and hardly smell the presence of objects beyond the water surface) the caretaker's opportunities for knowing and interacting with salmon are restricted. During feeding, for instance, salmon appear as brief repeated ruptures on the water surface, or as quick shadows moving just underneath, creating a "simmering" effect on the water. Occasionally, a few of them will jump, quickly exposing part of their body above water. Gradually, as the meal goes on and the salmon are less hungry, the simmering effect subsides. On salmon farms equipped at a low level of technological sophistication, these subtle changes are the only sensory information available that may indicate to him or her that the feeding should stop. However, their interpretation is difficult. Most salmon never come to the surface to feed but remain a few meters below and catch pellets as they slowly sink. Furthermore, the tendency for salmon to feed on the surface varies from one pen to another, and also depends on weather conditions. This example illustrates how, in the case of marine farming, the water surface presents itself as a boundary in relation to human sensory perception of the animal (and vice versa) and thus restricts the scope for "knowing the other" that constitutes the human–animal relation in most terrestrial domesticated contexts.

Transnational technologies of salmon farming offer a number of tools to help overcome such limitations. Most important in relation to feeding is the digital underwater camera that may be located at various depths inside the salmon pen. When hooked up to a monitor with a computer screen, this enables the farm hand to watch the activity inside the salmon pen while the salmon is being fed. Often, such cameras are used together with automatic feeders that may be turned off (manually or automatically) as a result of a specific density of pellets recorded on the screen. In this way, the lack of transparency across the water surface is overcome, and both salmon and pellets may be monitored, and mastered, at much greater levels of precision (Lien in press a).

As I have indicated, salmon farming involves a constant dialectic between monitoring fish behavior, interpreting available signs, and administering an adequate response. Overcoming the boundary represented by the water surface rests on the establishment of an interface, which in turn requires some kind of technoscientific mediation. For salmon farmers to come anywhere near the level of precision involved in terrestrial farming, a relatively sophisticated technoscientific regime is thus required. This makes salmon farming more deeply constituted by transnational networks of technology transfer than for instance the domestication of terrestrial species such as horses (Cassidy 2002), cows (Risan 2003), or guinea pigs (Archetti 1997).

In this section, I have discussed how the element of water makes a difference by introducing an element of distance in the human–animal relation. I have focused on the limits of the human sensory perception when engaging with species under water, and the interface established through technoscientific regimes. In the next section, I discuss the implications of domestication under water with a view both to the fluid qualities of water and to notions of property rights.

The Fluid Quality of Ocean Water

Discussing various technical and environmental challenges related to salmon farming, my Tasmanian informants often brought attention to what was in their opinion crucial differences between the conditions of a hatchery and the conditions at a marine grow-out site. In Tasmanian hatcheries, fry and young salmon swim in huge freshwater tanks in which practically all physical parameters are carefully monitored and controlled (O^2 levels, water temperature, bacteria levels, light, etc.). On marine farms, however, salmon swim in underwater pens or cages that are separated from the marine surroundings by a simple netting. This netting usually keeps the salmon in place, but is hardly sufficient to keep natural surroundings out.[6] This is especially important in relation to microorganisms, which spread easily with the water currents from one cage to the other, and even from one farm to the next, potentially spreading disease among salmon, and sometimes severe economic difficulties for the business enterprise.[7]

This difference between a hatchery and a marine site illustrates what is also a crucial distinction between terrestrial and marine farms, that is, the respective permeability of the boundaries that separate domesticated space from the surroundings. Clearly, terrestrial farm animals are also vulnerable to microorganisms that migrate by air (or to predators that break into their shelters). Yet, such challenges appear to multiply in marine contexts, simply because the fresh ocean water currents on which salmon depend, represent a continuous exchange of water and microorganisms between the domesticated contexts and their surroundings. Nature simply cannot be "locked out."

This has several implications. First, it disturbs the established conceptual distinction between the domestic and the wild in some interesting ways. Farmed salmon's proximity to ocean water gives rise to widely shared cultural connotations of purity and untouched non-humanized nature, which may be exploited for commercial purposes. In advertisements for Tasmanian Atlantic salmon, the alleged purity of the ocean water often underpins quality claims, attributing to the ocean space a certain symbolic value that allegedly "spills over" to the domesticated commodity.[8] Thus, because the conceived boundary between the "inside" and the "outside" of the domesticated space is not absolute, a certain overflow from the outside to the inside is conceivable and becomes intrinsic to the exchange value of the salmon itself.

Second, as they are immersed in ocean water, salmon pens also represent a constant threat to the ideal of keeping nature pure and untouched by human activity. In Tasmania, this gives rise to different types of conflicts between the general public and the salmon industry, from concerns about the loss of "amenities" of coastal properties that happen to have a view to a marine farm, to concerns about the environmental effects of sedimentation that builds up on the ocean floor beneath the salmon cages. In addition there is concern, as mentioned, that salmon might escape and threaten the ecological balance that involves local native species. Thus, a small salmon farm located on the coast of Tasmania enters, by default, a contested field of distinctions between wild–domestic and native–introduced. Not only does it appear in a region that is culturally classified as a "hotspot" of ecotourism, and where restorations of pristine pre-European landscapes is a community endeavor (Lien in press b). It also involves a farming technology in which the boundaries that seal off the domestic from the wild are hopelessly permeable and allow a number of challenging transgressions. Why is this so? Partly, this is simply because of water, a liquid element capable of flowing into all spaces available and of transmitting microorganisms and residue on its way. But these are not the only properties of water that make a difference. As Strang notes referring to water in the river Stour: "Its material qualities—its composition, its transmutability, reflectivity, fluidity and transparency—are inherent, but also responsive to context" (2004: 245). The context in this case involves the social, moral, and legal configurations that regulate the use and understanding of ocean water.

The expansion of terrestrial agriculture has gone hand in hand with growth of urban settlements and has been facilitated by a massive transfer of agricultural surplus from poor peripheries to centers that lay the foundations of what was to become centralized nation states (Hart 2004). A key element of this development is the notion of land as something that may be subject to private property regulations and transmitted along the relations of kinship descent. Ocean and coastal waters, on the contrary, are more often classified as communal property (Ballinger and Helmreich n.d.; Boissevain and Selwyn 2004; Pálsson 1991). Although legal regulations vary, most marine resources on the North Atlantic rim as well as in Australia are regulated through some notion of shared user rights or as common property at various levels of scale (community, state, nation, or humankind). Such notions prevent, to some extent, the commoditization of ocean water.

Consequently, when Tasmanian salmon farmers establish a marine farming unit, they do not *own* the water in the way a farmer may own a piece of land. Rather, they acquire (from state authorities) a lease that grants them the right to set up facilities in a specified part of the coastal area, for a specified length of time. This implies that they have no right to alter permanently the seafloor on which their site is located. More precisely, any temporary changes (such as sedimentation that builds up underneath the cages) are, by definition, an issue of public concern, and should be prevented as far as possible. If similar restrictions were placed on terrestrial farmers regarding their alterations of landscape and soil quality, it would greatly alter the conditions of agriculture in most parts of the world. In other words, as marine aquaculture is usually not accompanied by property rights to an area legally defined or set aside for domestication purposes once and for all, they are bound to occupy a narrow, ambiguous, and permeable physical space set off for aquaculture, located within, and sometimes diminishing, other people's access to the surrounding coastal waters. Thus, although marine sites cannot be owned, aquaculture leases may be seen to represent a form of closure of the designated coastal space (Apostle, McCay, and Mikalsen 2002; Pálsson 1991). This is another significant example of a "difference that makes a difference" as domestication takes place under water. The full implications of this for the social and commercial organization of marine farming lie beyond the scope of this chapter.

Beyond the Terrestrial Blueprint

Terrestrial domestication has provided the model for the way most cultures conceptualize human–animal relations. Similarly, in archeology and anthropology, the domestic has been established as a realm distinct from the wild through notions of physical and social proximity, and often by an extension of notions of kin (see Russell this volume). As Russell points out, the concept of domestication is thus often associated with taming, and with a notion of shared "domus," which refers back to the etymological roots of domestication (see also Wilson this volume). Furthermore, the conversion of animals to individual or collective property is central to the concept of domestication.

In the case of marine farming, most of these characteristic features do not apply. Although farmed salmon acquire certain behavioral traits not found in their wild relatives, they are hardly subject to taming. Furthermore, they are not conceptualized as kin in any sense, and their dependency on ocean currents effectively prevents the sharing of a human–animal *domus*. The only defining feature that clearly applies to marine domestication, and that makes farmed salmon distinct from wild salmon, is that of property: although farmed salmon can be owned, wild salmon cannot (common property rights regulate the right to fish, not the salmon themselves; cf. Pálsson 1991). Thus, from a socioeconomic perspective, a characteristic feature distinguishing salmon farming from much wild salmon capture is that the former is done for profit. The value of farmed salmon relies entirely on its anticipated exchange value, which in turn is continuously established through market transactions worldwide. This is also a differentiating feature of intensive aquaculture in relation to extensive (pond-based) aquaculture, in which the use value of the fish is usually also of some importance. If we accept Parry and Bloch's (1989) claim that a Western understanding of markets is underpinned by a sharp distinction between the domains of markets and morality, salmon farming enterprises are identified as an industrial, profit-maximizing enterprise, occupying the global market sphere, and, thus, likely to be the target of a more general critique of the ills of global capitalism (Beck 2000).[9]

In addition, salmon farming is easily seen as a threat to what is locally perceived as a natural order of things, including domesticated species. Although Western perceptions of nature tend to be associated with an environment unmarked by human activity, this connection between "nature" and "wild" is not absolute (see Suzuki this volume). This is particularly salient in the sphere of food related domestication where certain practices that have been carried out for generations have become "naturalized," whereas other practices, usually more recent, are seen as artificial (Lien and Nerlich 2004). Thus, certain pastoral landscapes associated with traditional patterns of domestication are seen as attractive and even natural.[10] This can hardly be explained except by reference to the passage of time. Over the time span of several generations, certain agricultural practices and their imprint on the landscapes have intrinsic to the physical surroundings, and thus appear as if "it was always there." With a history of less than three decades (and most places much less), marine farming of Atlantic salmon can hardly pass as something that "was always there." Physical signs of underwater activity, often only visible as black plastic circles on the ocean surface are bound to appear unfamiliar. In this way, the very novelty of marine aquaculture is likely to make it far more disturbing in relation to what we appreciate as "nature" or "pastoral landscapes" than many forms of terrestrial agriculture.

Domestication in the Making: Concluding Comments

Whereas the Neolithic revolution is understood largely through the interpretation of archeological and historical sources, the "aquacultural revolution" is current and ongoing. Although the full implications of the latter remain to be seen, the emergence of intensive marine farming is a significant transformation that calls for further analysis. Anthropology's cross-cultural approach to processes of domestication, and its methodological reliance on both contemporary ethnographies and archeological sources, make anthropology well equipped to deal with some of the issues raised by the current expansion of aquaculture. In this chapter, I have discussed some of the questions that may be raised through such a comparative perspective. I have demonstrated how the use of domestication as a comparative concept may illuminate continuities and differences between terrestrial and marine farming, and how it may allow us to situate contemporary transformations in a historical perspective. Just as the concept of domestication may shed light on contemporary aquaculture, I have argued that studies of aquaculture may also challenge—and lead us to reconsider—the concept of domestication. Modeled largely on terrestrial farming, the term domestication evokes meanings and association that are not necessarily found in marine farming. Considering one in light of the other may help us gain a better understanding of both.

Notes

1. Although hunter-gatherers captured, tamed, and raised animals, they could not control and confine the animals sufficiently to establish an isolated breeding population, which is a defining feature of domestication. Sedentary life thus appears to be a prerequisite for the controlled breeding that transforms animals from "tamed" to "domesticated" (Harris 1996: 454).

2. In this chapter, when referring to human–animal relations, I use the term "animal" as shorthand for nonhuman species in general (excluding species of plants) unless otherwise indicated. "Animal" thus includes terrestrial animals, birds, fish, and crustaceans.

3. The main investor was a company by the name of Selmer Sande, which was also involved in the Norwegian property market, which boomed in the mid- to late 1980s. Norwegian salmon farming regulations in the early 1980s put a restriction on the size of aquaculture enterprises in Norway. The general optimism in relation to aquaculture, combined with limited opportunities for growth of Norwegian enterprises, paved the way for such transnational expansion.

4. In some parts of Norway, salmon has historically been a significant and abundant food source for the local population but also a source of recreation for the urban elite.

5. "Feral" is very common term in Tasmania nature discourse, and is used about animals that survive without human protection in the Tasmanian wilderness, such as cats. Interestingly the term "feral" is only meaningful within certain environmental and cultural contexts. It cannot, for instance, be translated to Scandinavian languages, in which no such term exists.

6. The netting is, however, often too fragile to prevent native Australian fur seals from breaking into the cages, leaving major "loopholes" from which salmon may escape (see Lien 2005).

7. In Tasmania, the most problematic harmful microorganism in relation to salmon farming is amoebas that cause Amoebic Gill Disease (AGD). In the Northern Hemisphere, a different set of infectious viral and bacterial diseases occur, diseases that have not (yet) spread to Tasmania.

8. The website of the major Tasmanian salmon producer Tassal, states, for instance: "From the free flowing clean waters of Southern Tasmania comes one of nature's most exquisite indulgences … the salmon on your table today is as excellent as the environment it came from" (available at: http://www.tassal.com.au). The proximity to clear and clean oceans water is also referred to in the promotion of Atlantic Salmon from Northern Europe.

9. Although this applies to many terrestrial farmers as well, the latter may also bring the attention to affectionate relations, compassions, taming and

other examples of interspecies communication that may render their practice more acceptable in the eyes of the outsider.

10. Examples from Northern Europe would be sheep grazing in between stone fences in the green hills of Wales (the lack of trees and the manmade arrangements of stone are rarely seen as a problem), or the lush green grass outside a Norwegian mountain cottage (the altered soil conditions owing to prolonged presence of goats and sheep is rarely seen as a problem).

References

Apostle, Rocjard, Bonne McCay, and Knut H. Mikalsen. 2002. *Enclosing the commons.* St. Johns, Canada: ISER.

Archetti, Eduardo P. 1997. *Guinea pigs: Food symbol and conflict of knowledge in Ecuador.* Oxford: Berg.

Ballinger, Pamela, and Stefan Helmreich. n.d. *Recasting Maritime Anthropology.* Unpublished paper for the "Recasting Maritime Anthropology" panel, presented at the 101st Annual Meeting of the American Anthropological Association, New Orleans, November 20–24, 2002.

Beck, Ulrich 2000. *What is globalization?* Cambridge: Polity Press.

Boissevain, Jeremy, and Thomas Selwyn 2004. *Contesting the foreshore.* Amsterdam: Amsterdam University Press.

Cassidy, Rebecca. 2002. *The sport of kings; Kinship, class and thoroughbred breeding in Newmarket.* Cambridge: Cambridge University Press.

Clark, Nigel. 1999. Wild life: Ferality and the frontier with chaos. In *Quicksands: Foundational histories in Australia and Aotearoa New Zealand,* edited by Klaus Neumann, Nicholas Thomas, and Hilary Ericksen, 133–152. Sydney: University of New South Wales.

Clutton-Brock, Juliet. 1994. The unnatural world: Behavioural aspects of humans and animals in the process of domestication. In *Animals and human society: Changing perspectives,* edited by A. Manning and J. A. Serpell, 23–35. London: Routledge.

Douglas, Mary. 1975. *Implicit meanings.* London: Routledge.

Dunlap, Thomas R. 1999. *Nature and the English diaspora.* Cambridge: Cambridge University Press.

Ellen, Roy. 1994. Modes of subsistence: From hunting and gathering to agriculture and pastoralism. In *Companion encyclopedia of anthropology,* edited by T. Ingold, 162–197. London: Routledge.

FAO. 2003. *Some basic facts about aquaculture.* Rome: FAO. (available at: www.fao.org.docrep/003/x7156e/x7156e02.htm)

FAO Yearbook. 2001. *Yearbook of fishery statistics.* Summary Tables A–1, A–6. Rome: FAO Information Centre. Available at: http://www.fao.org/fi/statist/summtab/default.asp.

Gross, Mart R. 1998. One species with two biologies: Atlantic salmon *(Salmo salar)* in the wild and in aquaculture. *Canadian Journal of Fisheries and Aquatic Sciences* 55: 131–144.

Head, Lesley, and Pat Muir 2004. Nativeness, invasiveness and nation in Australian plants. *The Geographical Review* 94(2): 199–217.

Harris, David R. 1996. Domesticatory relationships of people, plants and animals. In *Redefining nature,* edited by Roy Ellen, 437–463. Oxford: Berg.

Hart, Keith. 2004. The political economy of food in an unequal world. In *The politics of food,* edited by Marianne E. Lien and Brigitte Nerlich, 199–220. Oxford: Berg.

Huntingford, Felicity A. 2004. Implications of domestication and rearing conditions for the behaviour of cultivated fishes. *Journal of Fish Biology* 65: 122–142.

Ingold, Tim. 2000. *The perception of the environment: Essays in livelihood, dwelling and skill.* London: Routledge.

Kaiser, Mathias. 1997. Fish-farming and the precautionary principle: Context and values in environmental science for policy. *Foundations for Science* 2: 307–341.

Leach, Helen. 2003. Human domestication reconsidered. *Current Anthropology* 44(3): 349–368.

Lever, Christopher 1996. *Naturalized fishes of the world.* San Diego: Academic Press.

Lien, Marianne E. 2005. King of fish or feral peril: Tasmanian Atlantic Salmon and the politics of belonging. *Society and Space* 23(5): 659–672.

——. in press a. Feeding fish efficiently: Skilled vision and universalising expertise. *Social Anthropology.*

——. in press b. Weeding Tasmanian bush: Biomigration and landscape imagery. In *Holding worlds together: Ethnographies of knowing and belonging,* edited Marianne E. Lien and Marit Melhuus. Oxford: Berghahn.

Lien, Marianne E., and Brigitte Nerlich, eds. 2004. *The politics of food.* Oxford: Berg.

Low, Tim. 199. *Feral future: The untold story of Australia's exotic invaders.* Melbourne: Penguin.

Macnaghten, Phil, and John Urry. 1998. *Contested natures.* London: Sage.

Montgomery, David R. 2003. *King of fish.* Boulder, CO: Westview Press.

Naylor, Rosalind L., Rebecca J. Goldburg, Jurgenne H. Primavera, Nils Kautsky, Malcolm C. M. Beveridge, Jason Clay, Carl Folke, Jane

Lubchenco, Harold Mooney, and Max Troell. 2000. Effect of aquaculture on world fish supplies. *Nature* 405(29): 1017–1024.

Pálsson, Gisli. 1991. *Coastal economies, cultural accounts.* Manchester: Manchester University Press.

Parry, J., and Bloch, M. 1989. *Money and the morality of exchange.* Cambridge: Cambridge University Press.

Reitz, Elizabeth, and Elizabeth S. Wing. 1999. *Zooarcheology.* Cambridge: Cambridge University Press.

Risan, Lars. 2003. Hva er ei ku? Norsk Rødt Fe som teknovitenskap og naturkultur [What is a cow? Norwegian Red Cattle as technoscience and nature-culture]. D.Art dissertation, Centre for Technology, Innovation and Culture, University of Oslo.

Rolls, Eric. 1969. *They all ran wild.* London: Angus and Robertson.

Russell, Nerissa. 2002. The wild side of animal domestication. *Society and Animals* 10(3): 285–302.

Stead, Selina M., and Lindsay Laird. 2002. *Handbook of salmon farming.* Chichester, U.K.: Springer-Praxis.

Strang, Veronica. 2004. *The meaning of water.* Oxford: Berg.

Trigger, D., and Jane Mulcock. 2005. Native vs exotic: Culture discourses about flora, fauna and belonging in Australia. *Sustainable Development and Planning* 2(2): 1301–1310.

Utter, Fred, and John Epifiano. 2002. Marine aquaculture: Genetic potentialities and pitfalls. *Reviews in Fish Biology and Fisheries* 12: 59–77.

Woods, Mark, and Paul V. Moriarty. 2001. Strangers in a strange land: The problem of exotic species. *Environmental Values* 10: 163–191.

Zeuner, Frederick E. 1963. *A history of domesticated animals.* New York: Harper and Row.

Putting the Lion out at Night: Domestication and the Taming of the Wild

Yuka Suzuki

Configuring the wild as drifting toward the edge of disappearance has become second nature in modern sensibilities. This imagined ephemeralness has worked itself into the very structure of our engagements with nature, shaping our desires and deepening our romance with it. In recent years, however, the idea of the endangered wild has been complicated by new forms of agrarian ranching that involve the production of wildlife. Seemingly contradictory by definition, the concept of "producing wildlife" upsets traditionally understood boundaries between the wild and the domestic, and public nature versus private property. In western Zimbabwe, a region of the world where former cattle ranchers have converted their properties into wildlife ranches over the past three decades, wildlife producers cater to photographic safari tourists and sport hunters who come to fulfill their fantasies of "wild Africa." In this case, the prosperity of the local economy depends on the ability to convince tourists of the irrefutable authenticity of its natural wild setting. Resident white farmers must therefore package and sell the wild, which is a remarkable feat, considering the fact that the "wild" landscapes of the region are actually less than 20 years old.

Because of its distinction from more traditional forms of nature preservation, this new type of production falls under the rubric of sustainable utilization, in which the categories between wildlife and game become much more fluid. Interestingly, in contrast to the transformations from game to wildlife that swept the world less than a century ago (Beinart and Coates 1995), the process is occurring once again in many places, only this time in reverse, turning wildlife back into game. The logic is by now familiar. The designation of a species for sustainable utilization wins the animal better odds in the long run (Duffy 2000). When the state officially classifies something as game, the animal in question enters into a new sphere of state monitoring, and becomes incorporated firmly under an umbrella of interests that differ from those of "straightforward" conservation. Under these auspices, the creature enjoys a new level of consideration based on the public recognition that it now has a deliberate purposefulness. Whether providing entertainment and sport, or putting meat on the table, the animal serves a concrete function; it is *utilitarian* in the literal sense. Game animals, moreover, evoke a permeable moral boundary, one in which management is perceived as being less problematic, and at times even imperative. In contrast, wildlife as an ideal continues to be defined by maintaining the illusion of nonintervention. Thus, as with other forms of domesticated property, game is more easily managed, quantified, and administered than wildlife, mainly through the practice of regulating populations, quotas, and hunting rights. In other words, unlike wildlife, which is identified through a conceptual segregation from culture, game is desirable for its manageability in terms of governance.

The confluence among issues of wildlife, game, domestication, and property has particular resonance in southern Africa, where the politics of wildlife and game have been foundational in the creation of postindependence national identities and economies. Alongside its neighbors, Zimbabwe has invested a tremendous amount of energy into building an elaborate national imagery around animals, and highlighting its wildlife policies as a cornerstone in claims to modern, democratic statehood. Rhodesia, as the country was known before independence, began to diverge from the beaten path in 1975, when the government passed the Parks and Wildlife Act that gave landowners the right to claim any wild animals found on their lands as their own private property. This act of legislation dramatically reinscribed the social role and meaning of wildlife—it became, all at once, something that was potentially profitable, rather than a constant headache that threatened other forms of agrarian production.

To convert wildlife into property was the equivalent of assigning animals with individual, quantifiable value; the next step, then, lay in building new structures that would enable the actual *application* of that value. Accordingly, in the late 1970s, Rhodesia challenged Western environmentalists by espousing sustainable utilization as a new alternative to conservation. This was a daring, audacious act, provocative to an extent that is difficult to imagine now, when the argument for utilization has attained a virtually unassailable status. Despite the onslaught of outrage and violent threats unleashed from around the world, the Rhodesian government refused to be deterred. Initial experimentation with the utilization paradigm resulted in the development of a program called Wildlife Industries New Development for All (WINDFALL), which was designed to distribute the benefits of wildlife resource use to residents of communal lands most affected by the presence of wild animals. Within a decade, WINDFALL gave way to the famed Communal Area Management Programme for Indigenous Resources (CAMPFIRE), which reinstated management rights over wildlife to communal areas bordering national parks. Weathering the storm through tumultuous controversy, CAMPFIRE since then has been both celebrated for its principles, and heavily critiqued for its failures in practice (Alexander and McGregor 2000; Dzingirai 2003; Hughes 2001; Madzudzo 1996; Murombedzi 2001; Murphree and Metcalfe 1997; Nabane 1994; Sibanda 2001).

What this program excelled at most, however, was in establishing a high profile platform for sustainable utilization. Within this framework, the commodification of wildlife occurred in three different ways. The first was through the culling of "surplus animals"—a term that represents a de facto shift to the vocabulary of markets—harvesting their skins, tusks, and meat, and selling these products for revenue; the second involved contracting with "safari operators" whose "clients" paid handsome fees for the trophy animals they hunted; and the third was based on bringing visitors into black communal areas for photographic safari tours. Of these, the third option proved to be the most logistically difficult, because the communal areas never had a consistent critical mass of wildlife, and the animals that were present were perceived by tourists to be "out of context" when situated against a backdrop of village houses and crops. Because of the nature of these particular applications, therefore, the commodification of wildlife in this context was synonymous with its transformation into game.

Given this broader historical shift in utilitarian approaches to wildlife, in this article, I explore the concept of domestication as a potential framework in rethinking the relationship between game and wildlife, the role of animals in the making of human social identities, and moral hierarchies attributed to the taming of nature. In the following sections, I briefly trace the gradual development of wildlife production in a small rural community in western Zimbabwe, fictively dubbed Mlilo. In the second half of the article, I turn to a closer examination of one particular species, or the cultural politics of lions, in identifying the delicate balance between domestication and the illusion of the wild that must be maintained in ensuring the economic, cultural, and political logics of wildlife production.

Cattle Conundrums

When white settlers first arrived in this region in the western part of the country at the end of the 19th century, they came with dreams of one day owning thousands of head of cattle and commanding positions of importance within Rhodesia's agrarian-based economy. However, what they found were dry, nutrient-deficient soils that made grazing difficult, and a stubborn weed that was poisonous to their livestock (Palmer 1977). In addition to having to cope with the highly volatile market for beef, over time, it became clear that the fundamental and greatest disadvantage for these ranchers lay in Mlilo's proximity to Hwange National Park and its resident animals. It was then that with the blessing of the state, which recognized the irreconcilable conflict of interest between wildlife and cattle presence, the settlers in the area embarked on systematic projects of wildlife eradication. This practice continued throughout the 1950s and 1960s, focusing on specific species—particularly buffalo—that were potential carriers of contagious diseases that could be transferred to cattle. Natural predators such as lions, cheetahs, leopards, and wild dogs were vilified, and consequently became the targets of sophisticated and vicious campaigns of extermination (Carruthers 1989; Mutwira 1989). Officially classifying these animals as "vermin," the government paid bounties for the skins of hyenas, jackals, tiger cats, lynx, mongoose, baboons, grey monkeys, cheetahs, leopards, and lions.[1] Aside from the utilitarian argument for eliminating game, there were significant moral stakes as well. Antigame rhetoric was the natural extension of certain Christian traditions that condemned, according to Matt Cartmill, the wildness of animals as "a satanically incited rebellion against man's divinely constituted authority over nature" (1993: 54). As a result, "disobedient" wild beasts have often been viewed as demons and sinners across time and space, symbolizing the fallen condition of humans in Christian thought. Clearly, in the eyes of the state as well as its citizenry, the game had to give way to the claims of crops and cattle for the civilizing project (Mutwira 1989).

Other factors, such as the liberation war in the 1970s, significantly increased the odds against cattle ranching. During this period, freedom fighters traveled in the bush and launched attacks on white farms in the area, making regular patrols to monitor the movement and safety of cattle virtually impossible. Under these circumstances, people had no choice but to leave their animals to fend for themselves. Incidents of stocktheft by freedom fighters proliferated, specifically targeting white-owned cattle of imported pedigree; other cows and bulls were shot or mutilated in symbolic defiance, and left to die within their pastures (Grundy and Miller 1979). In the Tribal Trust Lands, where the policy of compulsory cattle dipping had long been regarded with suspicion and resentment, the issue became radically politicized. Liberation war guerrillas encouraged villagers to destroy dip tanks by filling them with stones and lumps of concrete, while dip attendants risked death if they insisted on carrying out their jobs. Two years after the compulsory dipping ceased, tick-borne diseases began spreading like wildfire in the communal villages, and panicked villagers sold their cattle as quickly as they could get rid of them. As a result, prices in the overall cattle market plummeted, and created an even bleaker scenario for white ranchers. Thus, slowly but surely, the stage was set for a progressive disengagement from cattle, and the tide had ushered in a new age of wildlife renaissance.

Reenchanting Wildlife

The idea for wildlife ranching originally took root in 1975, when the establishment of the National Parks and Wild Life Act gave landowners the right to manage and benefit from wildlife found on private lands. The promulgation of this act emerged from a 15-year period of tentative institutional reform in attitudes toward wildlife, propelled by the Wild Life Conservation Act in 1961. Up until that point, wildlife numbers had suffered tremendous declines because of their perceived incompatibility with the country's development. It was only in the 1960s that the government came to the realization that even the animals within game reserves set aside specifically for their protection were falling under the threat of agricultural expansion (Child 1995). At that point, the administration recognized that the policy of centralized protectionism, in which animals had value only for the state, had become largely unenforceable. With the changing structures of governance brought by the Unilateral Declaration of Independence in 1965, the state adopted a deregulatory stance toward wildlife based on its inability to manage game on a national scale (Wildlife Producers' Association of Zimbabwe 1997). Consequently, the concept of extending the value of wildlife to individual landowners was formulated, with the objective of redefining animals as resources and forcing farmers to think twice before eradicating game resident on their properties.

Although the actual process of transformation to wildlife production in Mlilo was variably interpreted and bitterly contested at every turn, with the rising importance of the wildlife industry in the contemporary national landscape, many people today tend to forget their initial reluctance, and instead emphasize their roles in bringing about this historical shift. Piet Klaveren's story was one of the most beautifully crafted narratives that laid claims to pioneering the movement in the valley. Even back in his cattle ranching days, Piet was regarded by the rest of the community as somewhat suspect, a renegade with too many strange ideas that unsettled the traditionalists around him. Taking note of the exponential growth in the number of foreign tourists who visited Hwange National Park each year after Independence, Piet decided to turn Mlilo's proximity to the park—the very factor that had thus far been its downfall in cattle ranching—to his advantage, and transport the benefits of the tourist industry into the private sector. He set about performing the previously unimaginable by opening up his property to wildlife, the very antithesis of his conditioning as a cattle rancher for the past 20 years. Furthermore, he began establishing contacts with overseas hunters who have since become devoted clients, and built the first luxury lodge in Mlilo. Subsequently, his business has expanded to include a hunting camp, a backpackers' rest, and an upmarket bush camp in addition to the original lodge. The Klaveren family ultimately became one of the wealthiest and most influential in Mlilo largely because of his revolutionary visions, as well as the entrepreneurial savviness of his four children who eventually took over the business. Today, he never tires of telling the story of a U.S. ecologist he befriended during the 1980s, who pronounced that Piet was "a hundred years ahead of his time."

It was in the early years, however, before the benefits of his insurgent schemes became clear, that Piet encountered the most resistance from the community, which harbored a deep-seated distrust and dislike for wildlife. He began his efforts to entice wildlife onto his land by dismantling all of the fences that lay along the boundaries he shared with Hwange, and creating new water pans by pumping water from underground aquifers.[2] One of the more noteworthy landmarks on his property was a *vlei* that soon became a favorite roaming spot for a small herd of zebra.[3] Although this addition represented a triumph for Piet, his neighbors felt profoundly threatened by the new presence. From that point on, if any of the surrounding ranchers happened to find wildlife on their properties, they immediately assigned blame to Piet, regardless of the unestablished origin of the trespassers. Thus, although the zebra, impala, or wildebeest that raised an uproar could easily have come from the national park, as they had been known to do dismayingly often in the past, now people automatically assumed that they came from Piet's renegade ranch. For a period of several months, he received weekly invoices from a particularly cantankerous neighbor who insisted on charging him for "x wildlife species found grazing on my land on y date for z amount of time." In a heated physical encounter during a monthly meeting of the local Commercial Farmers Union chapter, the neighbor threatened to kill Piet if he discovered that his cattle had contracted foot-and-mouth disease from any kind of wildlife. Most of the farmers rallied around the neighbor, and Piet found himself stigmatized as the black sheep of the community for a period of time.

Despite the initially violent opposition to Piet's project, it was only a few years before others began to see the wisdom of his endeavors. Ranchers with more financial resources at their disposal began investing heavily in the acquisition of wildlife, as well as the construction of lodges and hunting camps to serve the growing number of people who chose Zimbabwe as a destination for their holiday trips. In other households lacking the capital to set up enterprises for themselves, the hunting concessions on their lands were leased to hunters who needed a place to bring their clientele. Alternatively, some people chose simply to lease portions of their property to lodge companies looking for picturesque sites on which to build what they hoped would be the next trendy hot spot in accommodation.

Thus, by the mid-1990s, everyone in Mlilo found themselves participating in the wildlife industry for their primary source of income. Those who were not actively involved in the procurement of animals hoped to reap benefits simply by opening up their properties, converting their cattle troughs into water pans that blended more aesthetically with the landscape, and sitting back to wait for wildlife to appear from their neighbors' properties or the national park. At the same time, though, the uniformity in dependence on wildlife belied a volatile terrain as people attempted to come to terms with the changing place of wildlife in their lives. The transition to wildlife ranching was by no means synonymous with people shifting their allegiances to an alternative worldview, or one that magically transported wild animals into the realm of the good. For the majority of the community, the choice was principally an economic one. As a consequence, the turn to embracing wildlife as a desirable presence in Mlilo was still a new and incomplete project by the end of the 20th century.

The Masquerading Bushpig

The turn from cattle ranching to wildlife production in Mlilo does not tell a tale of the unraveling of domestication; rather, it shows the gradual unfolding of a *new* type of domestication. Aside from its obvious economic benefits, wildlife production also had other affinities that made it all the more appealing for white farmers in Mlilo. The postcolonial state in recent years has poured enormous energy into representing white Zimbabweans as unwanted remnants of colonialism, building up an artillery of ammunition against their continued tenure in the nation. One of the common angles used to delegitimize their claims to equal citizenship and landowners' rights has focused on the idea of visible productivity on white commercial farms. In this context, properties devoted to wildlife prove particularly vulnerable to being constructed as a less legitimate form of land use. By the nature of its movements, wildlife will always be more difficult to pin down, therefore giving the illusion of empty, unproductive landscapes. White farmers thus strategically appropriate the label of "production" in constructing a frontline defense against the state's claims that farmers have allowed perfectly good land to stagnate.

As consummate strategists, farmers play up a second dimension of wildlife production—the links to conservation. The great majority of farmers adhere very closely to hunting quotas set each year by the Department of National Parks and Wildlife Management. In 1992, when the Hwange National Park administration ran out of funding in the middle of one of the worst droughts of the century, farmers in Mlilo diligently maintained their water pans, even when every single water source in Hwange had dried up. During those months, thousands of animals made their way into Mlilo to find water, and farmers claim that all of them would have died had it not been for the hard work of local white farmers. By invoking this particular spirit of conservation, wildlife production thus enables farmers to construct themselves as good citizens, rather than white ones. Countering stereotypes within public propaganda that identify white Zimbabweans as a racist, self-interested economic elite, wildlife ranchers present themselves as working wholeheartedly in the interests of the nation-state.

Through a similar logic, returning once again to the concept of domestication, wildlife production provides its practitioners with a position of moralism based on the ability to discipline and control nature. The power to exert influence over the physical environment is often identified as the dividing line between humans and animals, or even in some cases, "civilized" humans versus "primitive" ones. Successful domestication implies mastery, knowledge, rigor, and a strong willfulness, or arrogance of perspective. Together, these qualities attribute to domestication a particular kind of moral order, which in turn, ties in with ideas about competency in specific landscapes, or the ability to move comfortably within one's environment. One particular experience I had during the course of my fieldwork became definitive in terms of highlighting the moral dimensions of wildlife production and farmers' claims of belonging in this region.

On an azure-colored wintry afternoon in July of 2001, I sat on a hard sofa in the sparse, well-scrubbed home of Charles and Ella Murphy, an elderly white couple who lived near the tiny post office in the center of Mlilo. The atmosphere was palpably tense, and the teapot tray lay untouched on the table in front of us. The land invasions of white commercial farms that had begun in February 2000 had finally reached this region the month before, despite what had gradually grown into a comfortable complacency that here in the arid bush, far removed from the politics of the capital, they would be for the most part forgotten. For the past 16 months, in fact, farmers in this area had been relatively unhindered in going about business as usual on their lands. With the arrival of the war vets in June 2001, however, Mlilo's residents were rudely awakened from such naive expectations as the new land occupants quickly laid the foundations for thatch houses, created their own bus stops, and claimed formerly private roads for their own exclusive use. In each of these stages of land invasion, the local government officials not only turned a blind eye but actually offered infrastructural support, substantiating the popular belief that these land occupations were state-sanctioned despite their constitutional illegality.

My earlier conversation with Charles and Ella had been interrupted by an official from the Department of National Parks and Wildlife Management, who had come to investigate urgent complaints by the war vets that they had been harassed by lions at their campfire the night before. The National Parks officer had stopped in at the Murphys' house to report that they were now looking for an officially designated "Problem Animal."[4] In the brief heated argument that ensued, Charles declared vehemently that under no circumstances would he allow the Department of National Parks to shoot one of his lions, which by rights belonged to him as his private property.

Ritual subtleties in the performance of power between black and white Zimbabweans had changed dramatically since the land invasions began, and the officer's only reaction was to shrug dismissively as he led his team off onto the Murphys' farm. I sat waiting with Charles and Ella, who looked grim, and had given up any pretext of normal conversation. The National Parks team returned two hours later, and the officer, somewhat less confident this time, stated ruefully that they had failed to find any signs of lions around the war vets' campsite. Instead, what they had discovered were bushpig tracks, revealing the recent presence of a solitary wild pig. The officer quickly followed up in explanation that although not large in size, the grunts of a bushpig can sound similar to those of a lion, and the mistake was therefore perfectly understandable. The terrified war vets' fears had been calmed, and the potentially explosive action of tracking down and killing one of the Murphys' lions was temporarily deferred.

On the departure of the National Parks team, the tension of the encounter slowly evaporated, and Charles's taut expression transformed itself into a grin stretching from ear to ear. "Can you believe it?" He shook his head incredulously. "The war vets couldn't even tell a bushpig from a lion!" He laughed uproariously, and Ella joined in his revelry. Very soon a celebratory atmosphere filled the room, and the Murphys, whom I had always known to be very restrained and formal, were slapping their hands on the coffee table and wiping tears from the corners of their eyes.

The story quickly circulated around Mlilo, and farmers breathed a collective sigh of relief to find themselves once again on comfortable terrain. They chose to interpret the incident as evidence that the land occupiers were fundamentally out of place on their newly claimed lands—not just from the white farmers' perspective, but also from the occupiers' own point of view—if they had failed to identify even the most elementary differences between a bushpig and a lion. The story winsomely illustrated not only the war vets' obvious unfamiliarity and lack of ease with the local landscape but also their basic fear of it as an imagined space of "wild" animals and "untamed" bush. Thus, the ill-fated encounter between war vets and bushpig served to delegitimize in spectacular fashion the moral authority of the land issue that had been so carefully propagandized by the government. Here, adding fuel to already deeply ingrained practices of critiquing state corruption, was a local example that perfectly illuminated the fundamental contradictions and flaws of the land redistribution project.

It was in this context that the tale of the phantom lion grew in dimension as it was relayed with increasing hilarity among the farmers of the valley. Jokes revolving around landscape and (in)competency appear as a common theme in many societies, usually with clear subtext aimed at undermining a group's right of residence. The strategic appropriation of this case of mistaken identity therefore served as a way for people to contest their marginality within the national arena, and engage in the creative reconfiguration and discrediting of state agendas. After weeks of growing uncertainty augured by the arrival of the land invaders, the white farmers seized on this particular story as a safe, and yet powerful, means through which they could dismiss the government's blundering efforts as doomed to failure. Through this process, the state, ordinarily perceived as a menacing threat, was transformed, at least momentarily, into an object of laughter.[5]

The delight that the farmers took in this story was derived, moreover, from another equally important dimension: the fact that the bushpig had been mistaken for a lion. Lion imagery recurs regularly throughout farmers' stories, alternating between spheres in which they are labeled as awe-inspiring creatures for hunters and tourists, on the one hand, and despicable predators that prey on livestock, on the other hand. Not surprisingly, more than any other animal in Mlilo, the lion is powerfully evocative in its configurations of symbolic meaning and sentiment, always present as a critical point of reference for the formation of rural white Zimbabwean identity.

Within the larger context, lions occupy an arguably uncontested space as the most magnificent land carnivore in the contemporary Western imaginary. From the Lion King to the Lions of Tsavo,[6] popular culture has anthropomorphized the lion by gifting it with cunning, wisdom, nobility, blood thirst for vengeance, and sovereignty over the Animal Kingdom. Similar symbolisms can be found in varied temporal contexts, as in the use of lions in the Roman gladiatorial arena, and captive felines in the private menageries of 18th-century French aristocrats, which demonstrated wealth, power, and prestige on the part of their owners (Ritvo 1987). In China, dogs were once bred to resemble miniature lions, and roamed around imperial courts as a living symbol of the greatness of the Chinese empire (Tuan 1984). These evocative representations are typical of the role that many "supreme predators" take on in cultural cosmologies, as in the example of the jaguar in Aztec society, which stood as a metaphor for spiritual power and elite status (Saunders 1994), or the mythologies of the Inuit, for whom the polar bear carries powerful associations of male authority and prestige (D'Anglure 1994).

The emotional investment in a particular conceptualization of the lion's predatory nature was in fact so strong that for many years, studies revealing that lions were often opportunistic scavengers who exploited the kills of hyenas were met with indignation and resistance by popular audiences (Glickman 1995).

Putting the Lion out at Night

In Mlilo, lions have always been polyvocal in their symbolism, moving through a shifting mosaic of meanings across time. The point of departure in telling tales about lions begins with a particular origin myth, when white Rhodesians first arrived in the region at the beginning of the 20th century, and set out to transform it into the site of a successful cattle ranching industry. Among the many resident predators in the area, lions were especially notorious for their unflagging persistence in tracking down the *bomas* where cattle were sheltered, despite the painstaking lengths the farmers went to each night to move the location of these enclosures. Moreover, lions would hunt in groups, targeting structural weaknesses in the fences, and employing artful strategy to gain access to the precious cattle. Up until the 1970s, the losses brought by lions remained discouragingly constant, resulting in the harsh reality that most farmers in the area struggled each year simply to break even.[7]

Narratives recounting this time are interlaced with the language of hardship, but also with an undercurrent of romantic longing, as men recount how they spent countless nights guarding their livestock from the flatbeds of their pickup trucks, rifle in one hand, lantern in the other. The solitude of those deep nights, lying alert and watchful under the luminescent starlight, draws emotive power from the time-less mythology of human desire pitted against nature as an adversary. Skilled marksmanship was glorified, and the number of feline bodies a man could claim was transformed into an index of masculinity. White Rhodesian nationalism in the mid–20th century emphasized qualities of ruggedness and resourcefulness, epitomized by the spirit of the *voortrekker,* or frontier experience. The collectively glorified history of pioneers migrating northward from South Africa to conquer the unknown wilderness was thus the pride and essence of Rhodesian-ness. In this equation, lions were key, for their strategic identification as the most dangerous and cunning adversaries enabled the convenient metonymic leap from conquering *them* to symbolically conquering nature as a whole. In this context, the narrative of unending struggle between humans and lions—both usually coded male, for this was a gendered discourse of constructing masculinity—is incorporated into the ongoing reproduction of a "mythico-history" (Malkki 1995) that asserts entitlement to the land through the ethic of hard work and unflagging perseverance.

A key shift occurred in the 1970s, when people began turning their attention to the untapped potential of wildlife tourism. A gradual conceptual reconfiguration emerged during this period, in which wildlife was transformed from an abominable presence into one that could actually be favorable and desirable in the landscape. This was when Jon Van den Akker embarked on the previously unimaginable: the domestication of lions. Jon's inspiration came in the early 1970s, when he came across two lion cubs who had been orphaned when their mother was killed by a PAC team. Stirred by a combination of pity, curiosity, and whimsy, he brought the two cubs back home to his family. Despite initial protestations by his wife Marie, to the delight of the six children in the family, the lions became family pets. N'anga and Lwane enjoyed special privileges—unlike the family's dogs, who were not allowed to cross the threshold into the house and had to sleep on the verandah—they moved freely in and out as they pleased.[8] With a wry laugh, Marie recounted how they used to put the lions out at night, but inevitably, by the morning, they had found their way back in and were sprawled across the living room carpet.

During the height of the liberation war, N'anga sometimes accompanied Jon and his unit during their foot patrols in the bush. The image of a dozen armed white Rhodesian men dressed in drab military gear, walking through the bush with a fully grown adult lioness by their side is a startling one, as it is intended to be. "Having a lion with us gave us protection," Jon explained one afternoon, as we sat down to a lunch of brown bread, pickled beets, and potato salad. "Whenever people saw her, they ran. No terrorists would ever *shupha* us when she was around."[9] N'anga obviously contributed her services as a watchdog—or watchcat, rather—but her value went beyond simply decreasing their chances of being taken by surprise. One can easily imagine how her presence, and what must have appeared to be her own volition in accompanying these men, might have lent the unit a kind of power that bordered on the magical.

"We used her to interrogate them," Jon lowered his voice. "A black man can get beaten up, but threaten to put him in a cage with a lion, and he'll tell you anything. So we would take them to her pen, and threaten to lock them up with her. We knew she wouldn't touch them, but they didn't know that. After that, they told you everything. Exact locations of camps in Zambia, where their weapons were, whatever you wanted to know. You never saw anything like it. These guys, they've lived in the bush here all their lives, but they're shit scared of lions. I don't know why." As Jon shrugged, I tried to banish the vision of an interrogation from my mind. Troubling as it may be, we see here yet another example of how humans achieve power and authority through the strategic manipulation of animals. And with his statement, Jon conveyed the clear message that even though they had lived in this region all their lives, blacks could never master the environment in the same way that whites could.

In the 1980s, Jon expanded his vision and built a pen enclosure for 28 semidomesticated lions on his land. The lions were galvanized into action whenever a large tourist group descended on the famous safari lodge in the national park, and needed to be guaranteed a spectacular lion sighting. Jon trained the cats to follow his pickup truck, and after depositing a hunk of animal carcass in the middle of a scenic grassland, drove quietly away. The tourists would then "happen" on this huge group of lions feasting on a "fresh" kill, and their safari would culminate in the fulfillment of their fantasies of a quintessential encounter with "wild" Africa. Later, after the tourists had returned back to the lodge for cocktails at sunset, Jon would retrieve the lions and restore them to their pen. With the ferocious roar of these great cats resonating regularly on the property, the household achieved a legendary regional reputation among both white and black Zimbabweans as people who magically held lions at their beck and call.[10]

By the early 1990s, the project of converting wildlife into valuable capital was still incomplete in its ideological dimensions, but on the level of practice, almost everyone in the valley had entered the new industry in some form or another. Those who identified themselves as "enlightened," "modern," and "progressive," were the farmers who very early on had converted their properties entirely to wildlife ranching; and they in turn discounted as conservative and nonprogressive their neighbors who insisted on retaining part of their land for cattle, which represented a source of economic security in the event that the wildlife angle failed. In this atmosphere of contention, a series of stories began to circulate about one family whose members were rumored to be rampant and incorrigible "lion killers." Unlike many of the farmers in the valley, the Hallowells had no electricity and lived very modestly, and were labeled "primitive" and "backwards" for their staunch advocacy of cattle ranching, as well as their refusal to allow wildlife onto certain areas of their property.

When several dead lions were mysteriously discovered around the valley, the community demonized the family and accused them of maliciously tracking down and exterminating these lions by the darkness of night. The Hallowells were thus thought to represent a dangerous threat to the integrity of the community as a whole; as nonconformists and "deviants," they were particularly vulnerable in a situation that carried resonance with witchcraft accusations. They were taken to court by their neighbors, and eventually acquitted for lack of evidence, but the rumors were so pervasive that people in the capital 800 kilometers away knew of the infamous Hallowell "reputation." Consequently, in dramatic contrast to the cultural meaning of lions just three decades ago, the protection of lions in this context became an index of morality, modernity, and productive *Zimbabwean,* as opposed to *Rhodesian,* citizenship. Thus, on the one hand, the acceptance and strategic domestication of lions by the Van den Akker family was seen to demonstrate an engagement with modern, postcolonial sensibilities; whereas, on the other hand, the Hallowells' attitudes were derided as a defiant mark of antimodernity and the refusal to be swayed by contemporary global trends. Given that colonialism itself was understood to be a vehicle for the practice and dissemination of modernity, there is a certain irony to the fact that a "colonial mentality" toward wildlife is subsequently refigured as antimodern. Moral understandings surrounding animals are thus mercurial in nature, and in this context, serve as an important benchmark in differentiating the colonial from the postcolonial.

The Lion's Share

By exploring wildlife production through a specific focus on the cultural politics of lions, the objective of this article has been to explore the rubric of "domestication" as a form of discourse and practice in constructing racial hierarchies, white identities, and moral cosmologies. Although the form of ranching in western Zimbabwe may have changed over the past three decades, in local worldviews, the successful management of wildlife is equally predicated on the ability to discipline and control nature. In social constructions of difference, the power to domesticate the environment is used as a common trope in establishing the dividing line between the civilized and the primitive, the scientific and the irrational, and the enlightened versus the barbaric. Deeply rooted within Judeo-Christian theology, ideas of maintaining and controlling nature exist within a valence of abstract associations, including knowledge, wisdom, industriousness, determination, and godlike authority (Thomas 1983). Hence, the position farmers assume is a comfortably familiar one. Combined with the global status of conservation, these values inscribe a specific moral order on wildlife production, which, in turn, reinforces the argument for rights of tenure in one's settled environment. The original economic motivation is thus eclipsed, at least in the public sphere, by a cultural understanding of wildlife production that provides testament to the superior morality and pure intentions of whites. The beauty of this logic, however, would fail these farmers in the long run.

In the mid-1990s, Jon transferred the few domesticated lions that remained on his property to a large pen located near the family's lodge. There, when the lions roared, guests staying at the lodge would assume they were listening to wild lions, and experience a visceral thrill. In July 2001, the last lioness of the group—also called N'anga after her grandmother—miscarried at a very late stage in her pregnancy. Her cubs were stillborn, but her body refused to expel them and she was in obvious pain. As soon as the problem was discovered, Jon called a veterinarian from Hwange, who performed an operation to remove the dead cubs from N'anga's uterus. Afterward, Jon and I maintained a close vigil for several days, trying to coax her to eat with fresh zebra meat that had been set aside especially from a hunt. N'anga never recovered her strength, though, and quietly died a week later.

Jon was inconsolable. He refused to leave his room for the rest of the day, and when he finally emerged late in the evening, he looked haggard. He shrugged in response to my questioning glance. "I'm tired," his voice was hollow. "My heart's broken, Yuka." It seemed a strange thing to say. After all, N'anga had been caged for the last several years of her life, and Jon rarely saw her. Looking back now, though, I realize that Jon was referring at that moment to more than just N'anga's death. For him, N'anga had temporarily become a symbol of something much larger. In her death, she had evoked something deeper within him.

The land invasions that began in 2000 foreshadowed the end of a way of life for white farmers in Mlilo. The epoch of wildlife tourism, with its new wealth, prestige, and symbolic capital was drawing to a close as the existing moral order was suspended. In a few short months, thousands of animals in conservancies like Mlilo around the country were killed by war vets engaged in protest against white commercial farmers. Here, once again, animal bodies served as a metonymic tool for political critique, and through their deaths, represented the dramatic reordering of national agendas in which white farmers could no longer claim a place. Animals were utilized for their powerful symbolism, and their destruction became the physical manifestation of erasing, in one violent act, the moral and legal universes inscribed by white Zimbabweans. If domestication is a cultural project, than this, too, might be conceived as a form of domestication, with its fundamental transformation of wildlife in the interest of serving an explicit agenda. Thus, the poetics of animals, with all of the artistry and license employed in human experience, places them at the very heart of cultural politics, in which they profoundly reconfigure, mediate, and transform our lives. The time is perhaps long past due, then, for animals to reclaim the lion's share of ethnographic analysis in the quest to understand culture.

Notes

1. Similar to the bounty system on the U.S. frontier, the government offered a reward of up to £2 for every pair of lion ears brought in by a settler.

2. These steps coincided with the gradual process of selling cattle on the beef market without rebuilding the herd as Piet had done each year in the past.

3. Adopted from Afrikaans, a *vlei* is a low-lying marshy area that typically attracts wildlife because of the presence of extra water.

4. "Problem Animals" constitute a legal category that identifies individual animals reported as causing havoc within village communities. Problem Animal Control (PAC), a subdivision of the Department of National Parks, has the authority to relocate or exterminate the animals in question.

5. For a theoretical elaboration of the role of "laughter" in mediating relations between state and civil society, see Mbembe 2001 and Bayart 1993.

6. The story of the man-eating lions of Tsavo, which was popularized by the 1996 film, *The Ghost and the Darkness,* is based on real events in Kenya in 1898. During the construction of a railway bridge over the Tsavo River, two large male lions systematically killed and ate 40 railway workers, bringing construction to a halt. The lions were eventually shot and killed, but have been immortalized by the speculations they raised about the intelligence and intentionality of lions. Their bodies are on display at the Field Museum in Chicago today.

7. One family, for example, tallied their losses to lions at 127 head of cattle for a single year.

8. N'anga means "traditional healer" in chiShona, and Lwane is short for isilwane, meaning "wild animal" in siNdebele.

9. Adopted from siNdebele, "shupha" means "to bother."

10. Once the land invasions began, the Van den Akker farm was the last in the valley to be occupied. Jon maintains that this is because there are those who still believe that the property is protected by lions.

References

Alexander, Jocelyn, and Joann McGregor. 2000. Wildlife and politics: CAMPFIRE in Zimbabwe. *Development and Change* 31 (3): 605–627.

Bayart, Jean-Francois. 1993. *The state in Africa: The politics of the belly.* London: Longman.

Beinart, William, and Peter Coates. 1995. *Environment and history: The taming of nature in the USA and South Africa.* London: Routledge.

Carruthers, Jane. 1989. Creating a national park, 1910 to 1926. *Journal of Southern African Studies* 15 (2): 188–216.

Cartmill, Matt. 1993. *A view to a death in the morning: Hunting and nature through history.* Cambridge, MA: Harvard University Press.

Child, Graham. 1995. *Wildlife and people: The Zimbabwean success: How the conflict between animals and people became progress for both.* Harare: Wisdom Foundation.

D'Anglure, Bernard Saladin. 1994. Nanook, super-male: The polar bear in the imaginary space and social time of the Inuit of the Canadian Arctic. In *Signifying animals: Human meaning in the natural world,* edited and translated by Roy Willis, 178–195. London: Routledge.

Duffy, Rosaleen. 2000. *Killing for conservation: Wildlife policy in Zimbabwe.* Bloomington: Indiana University Press.

Dzingirai, Vupenyu. 2003. Campfire is not for Ndebele migrants: the impact of excluding outsiders from Campfire in the Zambezi Valley, Zimbabwe. *Journal of Southern African Studies* 29 (2): 445–459.

The Ghost and the Darkness. 1996. Stephen Hopkins, dir. 109 min. Constellation Entertainment. Beverly Hills.

Glickman, Stephen E. 1995. The spotted hyena from Aristotle to the Lion King: Reputation is everything. In *Humans and Other Animals,* edited by Arien Mack, 87–123. Columbus: The Ohio State University Press.

Grundy, Trevor, and Bernard Miller. 1979. *The farmer at war.* Salisbury, Rhodesia: Modern Farming Publications.

Hughes, David. 2001. Rezoned for business: How eco-tourism unlocked black farmland in eastern Zimbabwe. *Journal of Agrarian Change* 1 (4): 575–599.

Madzudzo, Elias. 1996. Producer communities in a community based wildlife management programme: A case study of Bulilimamamangwe and Tsholotsho districts. Harare: University of Zimbabwe Centre for Applied Social Sciences Occasional Paper Series.

Malkki, Liisa. 1995. *Purity and exile: Violence, memory, and national cosmology among Hutu refugees in Tanzania.* Chicago: University of Chicago Press.

Mbembe, Achille. 2001. *On the postcolony.* Berkeley: University of California Press.

Murombedzi, James. 2001. Committees, rights, costs and benefits: Natural resource stewardship and community benefits in Zimbabwe's Campfire programme. In *African wildlife and livelihoods: The promise and performance of community conservation,* edited by David Hulme and Marshall Murphree, 280–297. London: James Currey.

Murphree, Marshall, and Simon Metcalfe. 1997. Conservancy policy and the Campfire programme in Zimbabwe. Harare: University of Zimbabwe Centre for Applied Social Sciences Technical Paper Series.

Mutwira, Roben. 1989. Southern Rhodesian wildlife policy (1890–1953): A question of condoning game slaughter? *Journal of Southern African Studies* 15 (2): 250–262.

Nabane, Nontokozo. 1994. A gender sensitive analysis of a community based wildlife utilization initiative in Zimbabwe's Zambezi Valley. Harare: University of Zimbabwe Centre for Applied Social Sciences Occasional Paper Series.

Palmer, Robert. 1977. *Land and racial domination in Rhodesia.* Berkeley: University of California Press.

Ritvo, Harriet. 1987. *The animal estate: The English and other creatures in the Victorian age.* Cambridge, MA: Harvard University Press.

Saunders, Nicholas J. 1994. Tezcatlipoca: Jaguar metaphors and the Aztec mirror of nature. In *Signifying animals: Human meaning in the natural world,* edited and translated by Roy Willis, 159–177. London: Routledge.

Sibanda, Backson. 2001. *Wildlife and communities at the crossroads: Is Zimbabwe's Campfire the way forward?* Harare: Sapes Books.

Thomas, Keith. 1983. *Man and the natural world: A history of the modern sensibility.* New York: Pantheon.

Tuan, Yi-fu. 1984. *Dominance and affection: The making of pets.* New Haven, CT: Yale University Press.

Wildlife Producers' Association of Zimbabwe. 1997. *Commercial wildlife production in Zimbabwe.* Harare: Wildlife Producers Association.

Of Rice, Mammals, and Men: The Politics of "Wild" and "Domesticated" Species in Vietnam

Pamela D. McElwee

In the early 1990s, several new species of large mammal, previously unknown to science, were discovered in Vietnam. One was a new species of bovid, named by international biologists as the *saola (Pseudoryx nghetinhensis)* in 1992; the other new species were the giant muntjac *(Megamuntiacus vuquangensis)*, discovered in 1994, and the warty pig *(Sus bucculentus)*, rediscovered in 1997 after years of presumed extinction. The discovery of these new "wild" species was heralded by international conservation organizations, and as a result of all the attention, new biodiversity projects were established and international development funds for them were tapped throughout the country. More than $100 million in aid to conservation of wild animals and protected areas was pledged to Vietnam at this time.

At the same time as new money was going to protect these new "wild" species, a different kind of biodiversity threat was looming over Vietnam—the loss of agrodiversity, particularly local rice varieties or landraces. New hybrid and high yielding varieties (HYV), particularly from China, were dominating the Vietnamese market in the 1980s and 1990s, and as a result, use of local varieties was declining, particularly the indigenous "floating rice" found in the frequently flooded Mekong Delta. Despite the fact that this loss of agrodiversity might have an adverse affect on the millions of people in Vietnam who were rural rice farmers, not a single international conservation NGO working in the country ever advocated the need to protect *domesticated* biodiversity.

In this chapter, I examine these two case studies of the saola and floating rice as a way to explore ideas of domestication. Within, I argue that the discursive worlds between "wild" and "domesticated," between the new and the familiar, between species and varieties, and between flora and fauna, all have contributed to the unequal attention to the two cases. I conclude that the links between biodiversity, "undiscovered" species, and notions of "wild" nature have worked to obscure interest in the (also threatened) domesticated species found in farmer-managed landscapes.

"Wild" Mammals and "Lost" Worlds

For many decades, the country of Vietnam was associated with the tragedies of war. But in the 1990s, Vietnam suddenly became a site of international environmental interest, thanks to the discovery of the new mammals previously undescribed by science. Vu Quang Nature Reserve, a highland corner of Vietnam bordering Laos where the animals were found, became internationally known virtually overnight (see fig. 10.1). Coinciding with increasing worldwide interest in biodiversity conservation in the 1980s and 1990s, calls quickly went out for the immediate protection of the new species and all possible habitat. The new discoveries were seen as a chance for international conservation NGOs to begin to flex their muscles in Vietnam.

For many years, as Vietnam had built up a fledging protected-areas system from fragmented forests and regenerating war zones, Vietnamese biologists and a few Western scientists had been surveying the flora and fauna of the country in piecemeal fashion. But most scientists felt that Vietnam was not going to have a particularly impressive list of species after so much war and turmoil, and because the country was so small and densely populated. This was particularly the case for Vu Quang. Beginning in the late 1950s, a large logging operation employing several hundred people had extracted valuable hardwoods from these forests, and even the designation of the area as a protected watershed reserve in 1986 barely stopped the flow of logs from the area. Vu Quang was no one's idea of a biological gold mine.

But scientists dispatched to survey Vu Quang in 1992 had struck a rich vein. Up on the traverse of the Annamite Cordillera mountain range dividing Vietnam from Laos, the biologists found a new animal, the saola (see fig. 10.2). Vu Quang, and Vietnam by extension, was big news, and became even bigger with the later finds of more new animal species, all in the same area. Press releases poured out of this previously

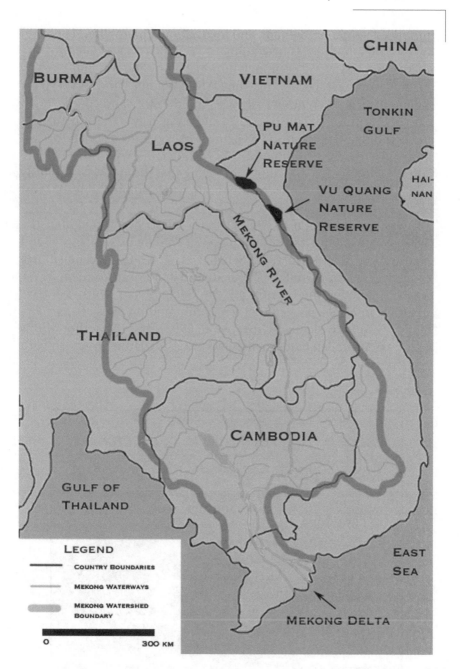

Figure 10.1. Map of Mekong Watershed, showing places mentioned in text.

Source: Base map from Mekong River Commission (available at: http://www.mrcmekong.org)

unnoticed corner of Vietnam, and international journalists clambered for a chance to see the new animals. "Scientists have described the reserve as a lost world seemingly untouched by the war and possibly teeming with new species," read one World Wildlife Fund (WWF) press statement. The *Washington Post* picked up the story and added, "With most of Indochina heavily populated and ravaged by wartime herbicides and bombing, stepping into Vu Quang is like opening a door into a lost and neglected place" (Briscoe 1992). Joshua Ginsberg of the Wildlife Conservation Society was quoted of saying that the area was "still truly wild" (Milius 1999). Almost without fail, the headlines all read, "Lost World Found" (*Time Magazine* 1992).

It was now common to find Vietnam—described throughout the 1980s by consultants as a dismal place for wildlife because of war and overhunting (Constable 1982; Schaller et al. 1990)—acclaimed as a "zoologist's dream" (Drollette 2000). In its handbook to Vietnam, WWF began to describe the country as "the Galapagos of Southeast Asia." An article in *Nature* posited that Vietnam's mountainous center was a "biotically unique region" (Groves et al. 1997), and the author of this report, an Australian biodiversity specialist, added in an interview with

Figure 10.2. Vietnamese stamp issued in 2000, with World Wildlife Fund support, featuring the *saola.*

a journalist that "Vietnam is the place to be since all the discoveries" (Drollette 2000: 16). The popular press universally proclaimed that Vietnam was now a "biodiversity hotspot" on the world stage (Drollette 1999; Tangley 1997).

The discovery of the new species was especially surprising given Vietnam's recent tragic history of war, strife, and authoritarian high socialism. Immediately after the discovery of the saola, WWF and Ministry of Forestry officials pledged to fund missions to find a live animal and to uncover crucial details about its distribution and habitats. The mission was to map this "lost world" and its scientific wonders. The lost world metaphor proved particularly strong in galvanizing international attention and international conservation dollars for Vietnam, as this small corner of the globe was envisioned as a particularly rich "living laboratory" for the protection and study of biodiversity (Linden 1994). As a result of all the international attention, the Vu Quang Nature Reserve was created from the former logging operation, and the acreage of the preserve was expanded from 19,000 to 54,000 hectares. Foreign money to run the park also began to flow, in the form of an $8 million grant from the Netherlands to WWF to form a "Vu Quang Conservation Project" (VQCP).

Vu Quang was not the only site of interest. Partially as a result of the new species, conservation came to Vietnam in a big way in the 1990s. NGOs tripped over themselves to set up offices in Hanoi; new biodiversity projects were established, and international development funds for them were tapped, not just in Vu Quang, but throughout the country. One of the more comprehensive of these international efforts was a preparatory study for the Biodiversity Action Plan for Vietnam (Socialist Republic of Vietnam 1995). Another was a large joint project of BirdLife International and various Vietnamese state agencies, funded by the European Union and other donors, to develop a comprehensive protected area system for Vietnam's biodiversity (Wege, Long, and Mai Ky Vinh 1999).

But the "lost world" could not be saved so immediately as the NGOs had hoped. In fact, the international attention brought many unwanted consequences as well. Westerners, from news agencies to self-professed "conservationists," were known to be offering large sums of cash for a live saola so a picture of it could be flashed around the world; 21 saolas were killed and three were removed alive to Hanoi between 1992 and 1994 (Schaller 1998). This number exceeded by far the number of saola skulls found in hunters' homes during the original research mission in 1992. A WWF report later that the "world-wide publicity campaign

... succeeded in mystifying the site but virtually arrested conservation efforts as the attention in the region led to competition among trophy hunters and also rendered access to the reserve difficult for researchers." The report went on to portray a "paralysis of conservation activities" and decried the fact that "the "lost world" of Vu Quang may once again sink into oblivion if steps are not taken to assure that Vu Quang remains central to Vietnam's conservation agenda" (Eve 1998: 1–2).

Yet, despite the flurry of attention that had focused on Vu Quang in the early 1990s, when the VQCP was at the end of its 5-year run, it was widely considered to have been a failure as a project and had produced no seeming reduction in the amount of wildlife taken from Vu Quang. An EU funded project in the Pu Mat Nature Reserve in the nearby province of Nghe An, where the saola was also later found, admitted in 2004 that their $27 million pledge had largely had little effect and that the saola population there had declined by 50 percent since the beginning of the project (Hardcastle et al. 2004). Vu Quang has now largely been left on its own, and a recent conference on the fate of the saola indicated that it is in more danger of extinction now than it was 10 years ago (Hardcastle et al. 2004).

The Vu Quang example was perhaps the most extreme case of the failures of biodiversity conservation in Vietnam, but many similar problems seemed to be echoed across the country. A number of articles in the Vietnamese press in the late 1990s highlighted the failures of biodiversity conservation across the board, because of high-level corruption in provincial forest protection departments, complicit involvement in the wildlife trade by customs, border guards, and the army, and confusing policy directions from the central government (Cao Hung 1999; Le Huan 1999). By 1999, one well-known Western conservationist, John Terborgh, even recommended that international organizations "write off" Vietnam as a "lost cause" for biodiversity (Terborgh 1999: 184). The optimism for biodiversity conservation that had characterized the early 1990s had dissipated by the turn of the millennium; one of the most well-known foreign conservationists who had worked in Vietnam since the early 1980s said to me at the end of my fieldwork in Vietnam in 2000, "I came to work in paradise, but paradise has been lost."

"Domesticated" Rice and "Familiar" Worlds

At the same time that the new species were being lauded and new measures being drawn up to protect them in the north of Vietnam, new environmental problems were emerging in the south, in particular the seeming rise in the destructiveness of the annual flooding cycle of the Mekong River. The Mekong is one of the world's greatest river systems, beginning in the Tibetan plateau of China and winding its way down to the Gulf of Thailand (see fig. 10.1). The Mekong Delta is the end point for the river and encompasses many of the most productive agricultural lands in mainland Southeast Asia. Most of the area is characterized by an altitude of less than 100 meters. The delta is also by far the most densely populated part of the entire Mekong basin, with average densities ranging around 500 people per square kilometer.

The mean rainfall for the lower Mekong basin ranges 1,600–2,400 millimeters per year, and the rainy six-month monsoon season from June to November can contribute as much as 90 percent of this yearly rainfall (Volker 1993). These rainfall rates have a dramatic effect on the flow rates of the Mekong and the difference between months can be on the order of 15 times magnitude (U.S. Engineer Agency for Resources Inventories 1968). The mean annual discharge of the Mekong has been estimated at around 475×10^9 cubic meters annually, placing the Mekong sixth in the world among river system water discharge rates (Pantulu 1986). There are no mainstream storage structures or dams on the Mekong to affect flow, and only a small number of tributary dams.

As might be expected with a free-flowing river with the rate of water discharge of the Mekong, flooding is common. Floods in the lower Mekong can last between 2 and 6 months, with 12–14 million hectares of land in the lower basin inundated each year (Daming 1995). The effect of these floods on life in the Mekong basin are varied. Although the floods are largely predictable in the seasons in which they occur, total levels of flooding are difficult to forecast. The Mekong Committee Secretariat described in 1988 the problems encountered by Mekong flooding:

> In broad terms, about half the delta is subject to flooding during the rainy season for 4–6 months and only low-yielding floating rice can be cultivated. Floods are relatively regular and gentle, and people have tuned their lives and agricultural practices to them. ... More productive rice varieties can only be cultivated if and where flooding is sufficiently reduced, but complete regulation of the flood flows by mainstream and tributary reservoirs is not a realistic perspective. (Jacobs 1996: 12)

The people of the lower Mekong have long had a saying that they have over time learned to "live together with the floods [*song chung voi lu lut*]" (McElwee and Horowitz 1999).

The Mekong Delta, despite the unpredictable floods, is enormously productive agriculturally. The delta contains only 25 percent of the agricultural lands of Vietnam (and only 11% of the total land area), but produces almost 50 percent of the nation's rice output (Nguyen Thi Song An 1996). Rice cultivation is the most important agricultural activity in the Mekong basin, and has been for hundreds of years (Hickey 1964). Vietnam suffered chronic deficiencies of rice during the Vietnam War and in the years following reunification, when much of the Delta was collectivized into inefficient production units. Yet the liberalization of the Vietnamese economy in recent years has allowed the Mekong Delta to become not only self-sufficient in rice but for Vietnam to become the third largest rice exporter in the world after the United States and Thailand (Pingali et al. 1997).

This jump from rice importer to rice exporter has been attributed to several factors. The first is the liberalizing of the economy and the shift to household production away from collectives (Pingali and Vo Tong Xuan 1992). The second is the introduction of new and improved "green revolution" HYV of rice, varieties that produce higher yields primarily through selective breeding of favored traits in seeds and the increased use of inputs of fertilizer and pesticides. These varieties also require irrigation and regular supplies of water and are not usually tolerant of flooding. These HYVs have nearly extirpated the use of local traditional varieties of rice in many parts of the Delta. But the loss of these traditional varieties may place many areas of the Delta at great risk of crop losses should Mekong floods continue to remain unpredictable, and the situation may be worsened by rising ocean levels caused by climate change. At the same time, however, no conservation NGOs working in Vietnam have sounded the alarm about the loss of Vietnam's agrodiversity. They have all remained silent on this domestic (and domesticated) issue.

Traditional Rice and HYV Rice

O. sativa is the rice species that predominates in Vietnam, and there are a huge number of land races and varieties within this species. As early as the 18th century, the well-known writer and historian Le Quy Don described hundreds of local cultivars of rice (Dap 1980). Can Tho University in the Delta has collected and stored more than 1,000 local rice varieties (Nguyen Huu Chiem 1994). There may well be many more; a recent study in nearby Laos documented over 3,100 distinctly named varieties of rice (Rao et al. 2002). Traditional varieties can be distinguished by farmers among various favorable characteristics, such as growing period, time of harvest, ability to be transplanted, and taste and physical characteristics such as whiteness of grain, solidness of the grain head, and so forth. Traditionally, early maturing rice was usually planted in coastal areas and broad depressions, where they could be harvested before saline intrusions in November (see Table 1). Medium varieties were often planted on relatively high areas of tide-affected flood plain where it is difficult to control water. Late varieties were cultivated in low lying areas with plentiful fresh water, and so-called floating rice was cultivated in the high floodplains (Nguyen Huu Chiem 1994). Each variety was chosen by farmers according to their experience in their local topography, and most farmers planted a range of varieties in any one season. If bad omens indicated floods or drought might occur, farmers would change to different varieties from year to year. Anticipated rice prices for the year also influenced farmers' decisions about what, how much, and when to plant.

Perhaps the most interesting of the traditional rice cultivars grown in the Delta are the aforementioned "floating" rice varieties, which have the unusual ability to extend their stalk rapidly during the flooding that accompanies the monsoon, so that the panicles of rice rest on the crest of the water until the flooding subsides. These varieties are uniquely adapted to flooding, as they elongate their internodes directly in accordance with water levels, and can be grown in water up to 3.5 meters deep. As river levels can rise 10–20 cm a day in the late rainy season in the Delta, the only crop that can be grown in these circumstances (without flood control) is this floating rice. Most HYV are not adapted to this sudden inundation and flooding, as the HYV are not photosensitive (i.e., dependent on the length of the day to

Table 1. Main Types of Traditional Rice Varieties in the Mekong Delta

Rice Variety	Duration in seed bed (days)	Growing period (incl. seed bed) in days	Time of harvest
Early varieties	30–40	130–160	Variable, Oct.–Nov.
Medium varieties	40–50	160–200	Dec.–Jan.
Late varieties	50–60	200–250	Jan.–Feb.
Floating	none	200–240	Dec.–Jan.

Source: Nguyen Huu Chiem 1994

determine growing patterns) and have a shortened maturation period; this is precisely the time when floating rice is elongating with water levels (Oka 1975). The flooding period also serves to keep weeds down without the use of herbicides, as the weeds drown while the floating rice continues to grow. The rice can then be harvested by boat before the floods recede, or afterward, as the rice dries in a tangled flat mass of stems on the bottom of the paddy field (Cummings 1978).

Floating rice, although classified as *O. sativa,* is believed to be more closely related to the wild ancestor of rice, *O. perennis,* as many *O. perennis* varieties exhibit floating tendencies (Oka 1975). Unimproved deepwater and floating varieties thus have more "natural" genetic variability than the improved modern varieties (MV) used in most parts of the world. Precisely because these deepwater varieties are adaptable to diverse environments and changes, their natural genetic variability is a positive. In the 1930s and 1940s, French agronomists considered these varieties to have a great potential to expand agriculture in the Mekong Delta, and they brought a number of Cambodian floating varieties to Vietnam to be planted (Pingali and Vo Tong Xuan 1992). However, although the floating rice is highly adaptive to the hydrological balance of the area, it is unfortunately not highly productive. Yields of floating rice are often relatively low, around 1.5 tons per hectare (Dao The Tuan 1985). Floating rice also matures in a longer growth period than other varieties, meaning that only one rice crop a year can be grown (see Table 1), although this can be supplemented with a dry winter season vegetable crop.

However, in 1967, at the height of the Vietnam War, improved HYVs were introduced to the Mekong Delta from the International Rice Research Institute (IRRI) in the Philippines. The introduction was partly political, as the U.S. government wanted to increase the number of rice farmers in the Delta who were sympathetic to their cause, and they hoped that increased rice production would be the incentive they could use to entice the farmers to their side (Logan 1971). The first MV introduced was a semidwarf variety called IR8. In Vietnam, the rice was called *Lua Than Nong,* or "Rice of the Farming God." Farmers also took to calling it *Lua Honda,* as one good crop of the IR8 could bring enough money to buy a Honda motorbike (Hargrove 2002). During the U.S. presence in South Vietnam, USAID set a high priority on introducing these HYVs throughout the delta and to gradually eliminate the "unproductive" strains like floating rice (Cummings 1978). Five years after they were first introduced, by the 1973–74 cropping season, more than one-third of the Mekong Delta's total rice area was planted to these MVs (Vo Tong Xuan 1995). During the war, the communist forces of the National Liberation Front (also known colloquially as the Viet Cong) had initially been wary of the new HYVs and their introduction by the Americans; they spread rumors that the IR8 "was a plot of the U.S. imperialists to spread leprosy and sterility" (Hargrove 2002: 77). Later, however, the NLF saw their potential and directed operatives to steal the seeds from U.S.-controlled areas and to plant it in "liberated" areas to gain the support of farmers.

After the unification of North and South in 1975, state emphasis was immediately on preventing postwar starvation by expanding rice production areas through reclamation and irrigation projects, and to move to collectivized agriculture. Northern agricultural cadres from Hanoi and the Red River Delta were set down to the Mekong Delta to administrate these new policies for cooperativization and agricultural expansion. The cadres were unfamiliar with floating rice, as it is not found or grown in the north of Vietnam, and they were adamant that this rice needed to be replaced by more productive varieties. Their plan was to replace the floating rice with HYVs, to speed up irrigation at the end of the dry season so that a spring rice crop could be planted, to plant a fall crop immediately after the spring crop while the waters were still high, and then let the annual flood waters overflow the dikes after the harvest of this fall rice (Nguyen Khac Vien 1985). The cropping pattern was not possible until the introduction of the MVs, with their shorter time to maturity. This complicated system of two rice crops grown successively to beat the floods required that irrigation canals be built throughout the floating rice region, as well as a new system of protective dikes that could convert the deepwater rice areas to irrigated zones. A number of pumping stations were built by the government, and many areas of the floating rice region were slowly converted to the master plan of doubled cropped rice. Since 1983, more than 300,000 hectares of floating rice land has been converted to double-cropped land through improvements in irrigation, decreasing the deepwater rice-growing areas from their prewar peak of 1.26 million hectares, or nearly half the Delta's total land area (Vo Tong Xuan et al. 1995).

Since IR8 first came to Vietnam, nearly 100 other improved MVs have been introduced, mainly from IRRI and Vietnamese research stations (Denning and Vo Tong Xuan 1995). Each new variety of rice with shorter cultivation cycles (now many with less than 100 days) has led to an explosion of double- and even triple-cropping cycles in much of the Mekong Delta (see Table 2). And the changes in types of varieties grown in Vietnam show even more dramatic changes (see Table 3). Whereas only 17 percent of the rice grown in Vietnam in 1980 was improved, hybrid, or "modern" rice, by 2000 the total was more than 94 percent (Tran Thi Ut and Kajiska 2006). HYVs currently predominate in all areas of Vietnam, from a low of 77 percent of total rice production coming from MVs in the mountainous Central Highlands to near total adoption (99.5% MVs) in the Mekong Delta (Tran Thi Ut and Kajisa 2006). Although some farmers have been hesitant to buy hybrid seed in particular, as they were wary of not being able to select and retain seeds for use the following year (thereby contributing to genetic improvement over time, a practice known in Vietnamese as *de lai*), the vast majority of the rice varieties now planted in Vietnam are MVs developed elsewhere in Vietnam or else imported from China or IRRI.[1] Overall, yields of MVs have outpaced traditional cultivars in all areas of the country (see Table 4). Triple cropping has raised the annual production of rice to an unbelievable 12 tons per hectare in some areas (Vo Tong Xuan

Table 2. Percentage of Rice Land under Cropping Systems, Mekong Delta

	1980 (%)	1990 (%)
Triple Cropped	1	4
Double Cropped	29	54
Single Cropped	70	43

Source: Nguyen Huu Chiem 1994

Table 3. Adoption of 'Modern' Rice Varieties (MV) in Regions of Vietnam

Year	Red River Delta (% of rice land area planted in MV)	Mekong Delta (% of rice land area planted in MV)
1980	52.9	9.7
1985	68.4	26.4
1990	78.5	48.3
1995	90.5	79.8
1998	92.2	87.7

Source: Tran Thi Ut 2002

Table 4. Yields of Improved "Modern" Rice (MV) and Traditional Varieties (TV) in Delta Regions, in Average Tons per Hectare

Year	Red River Delta		Mekong Delta	
	MV	TV	MV	TV
1980	2.6	2.0	2.5	2.3
1985	3.2	2.3	2.9	3.1
1990	3.8	2.1	4,0	3.4
1995	4.7	2.0	4.2	3.3
1998	5.3	3.1	4.3	2.8

Source: Tran Thi Ut 2002

and Matsui 1998), although this has come with great capital costs and the doubling of labor input for many farmers (Hossain, Duong Ngoc Thanh, and Gascon 1995).

The Effects of HYV Expansion

The very few areas with farmers that still grow floating rice have tried to innovate and keep up with the triple-cropped areas by expanding their winter dry season crops, like sesame, peanuts, and beans; in fact, this dual system has been shown to be as economically profitable as double-cropping HYV rice (Vo Tong Xuan et al. 1995). In fact, HYVs have pushed many farmers into less diverse overall economic strategies, as well as decreasing diversity of rice varieties, as HYV requires many more upfront capital and labor commitments than floating rice (Hoang Tuyet Minh 2000). Studies have shown that the share of income from rice in floating rice cultivating families is only 30 percent of total income (the rest being contributed by other crops, fishing and livestock, wage labor, trade, etc.), whereas 65 percent of the income of HYV irrigated rice producers came just from rice production (Hossain, Duong Ngoc Thanh, and Gascon 1995).

Yet these HYV are not suited to the flooding season of many areas of the Mekong delta, and they are particularly vulnerable to "exceptional event" flooding when dikes are overrun and bunds breached. An early monsoon or an extended monsoon can wipe out a HYV crop completely. The 1990s saw several major "over" floods during the monsoon, which resulted in millions of dollars of damage to crops and homes, and hundreds of deaths. In the past, when farmers had a wide variety of traditional varieties of rice to choose from, these flooding events were less damaging. As even the World Bank has noted,

> To counter this unpredictable rainfall and irregular flooding, the farmer relies on a contingency approach to planting schedules and methods of cultivation and uses a broad spectrum of rice varieties, each with special characteristics such as higher drought resistance, better tolerance of submersion time, better adaptability to the acid-sulfate sandy soils, or the ability to grow with the floods (floating rice). (World Bank 1994)

Should modern farmers be locked into agricultural schemes that do not allow for this last-minute flexibility (such as late HYV that waterlog easily or inflexible irrigation schedules), they will likely suffer badly. Clearly, the use of MVs are increasing some risk for farmers in years of extreme flooding and in more flood-prone areas.

Vietnam is not the only country that has experienced change induced by the switch to MVs. Nearly all the rice in China is now either HYV or hybrids; 5 million hectares of rice land in Thailand that used to be planted to multiple varieties is now planted in just three kinds (Rerkasem and Rerkasem 2005). Over 80 percent of the Philippines' rice production is now MVs as well (Morin et al. 2002). In each of these countries, research has shown the loss of rice diversity and variety that has accompanied the spread of these MVs. In other countries of Asia, like Cambodia and Laos, only a small amount of the total land area is irrigated, and traditional varieties have persisted longer in these areas as a result (Rerkasem and Rerkasem 2005).

For reasons of space, I do not explore in this chapter the problems associated with the Green Revolution and the new MVs, which have been well noted elsewhere (DeKoninck 1979; Dove and Kammen 1997; Farmer 1977, 1986; Jacoby 1972; Miller 1977; Yapa 1993). Proponents have argued that MVs from the Green Revolution have "succeeded in raising the health status of 32 to 42 million preschool children" over the last 40 years (Evenson and Gollin 2003), and that switching to higher yielding MVs has reduced "the pressure to open up more fragile environments such as uplands and tidal wetlands for rice cultivation" (Khush 1995: 280). Yet others have shown the use of MVs has had serious environmental problems, such as pesticide overuse—leading to both human poisonings and effects on natural animal life—along with fertilizer runoff, which damages lakes by causing eutrophication (Pimentel and Pimentel 1990; Singh 2000). Within Vietnam in particular, the role of HYV in increasing environmental damage has been noted, such as reduced fish production in paddies and increased pest infestations (Berg 2001; Heong, Escalada, and Mai 1994). This is caused by the increased dependence on pesticides by HYV farmers; surveys indicate that nearly 100 percent of HYV farmers in the Delta used pesticides on their crops, whereas only 45 percent of the farmers of traditional floating rice used them (Hossain, Duong Ngoc Thanh, and Gascon 1995). Other authors have also stressed that the change from floating to irrigated rice has increased income stratification, as poorer farmers have been unable to make the necessary investments in new inputs required by HYVs (Cummings 1978; David and Otsuka 1994).

The Two Cases in Comparison

What do these two case studies of the saola and the floating rice tell us about domestication? I believe they provide interesting bookends to discussions of what constitutes "naturalness" and "wildness," traits normally often juxtaposed with "domestication" and "tameness," as the other authors in this volume have amply demonstrated. What I want to emphasize here is that widespread associations of the contrasting traits of "domestic" and "wild" have very real and very serious consequences in the world today, none more so than in discussions of environmental conservation and biodiversity. We can see this problem extremely clearly in Vietnam. In all the hubbub about *biodiversity* that came to light in Vietnam in the aftermath of the discovery of the saola—no mention was ever made by any major conservation organization of Vietnam's decreasing *agrodiversity*. No conservation NGOs attended a 2000 conference on "In-situ Conservation of Native Landraces and Their Wild Relatives" sponsored by the Agriculture Ministry in Vietnam (see Institute of Agricultural Genetics Vietnam 2000). Nor have any major conservation groups taken a public stance on agrodiversity nor put any funding or attention toward this issue. They remain silent on the question of HYV and the loss of traditional rice cultivars. My question is: Why? Why did the many "green" groups that sprang up in Vietnam in the 1990s not include attention to agricultural biodiversity in their million dollar fundraising plans for Vietnam, and instead chose to focus solely on the biodiversity of places like Vu Quang's "lost world"?

This is a question I have tried to answer elsewhere with an ethnographic look at the people involved in biodiversity protection in Vietnam (McElwee 2003). International conservationists who spoke to me during my fieldwork in Vietnam often voiced their ideas about biodiversity in moralistic discourse about "saving nature," and often emphasized Vietnam's role in preserving biodiversity for the "world's heritage." One particular focus of international NGOs and their workers, particularly since the discovery of the new mammal species in the 1990s, was on procuring Vietnam's place in the world pantheon of biodiversity; project documents for biodiversity plans invariably referred to the newly discovered mammal species and noted that Vietnam was now a "world hotspot" of biodiversity. These international workers often moved from country to country and had access to materials that allowed them to compare biodiversity among regions. This has tended to focus some international conservationists' attention on specific unusual, rare, and/ or endemic species (such as the saola). Rice, being a global crop, was not something unique to Vietnam, but the saola was.

Another focus of the conservationists was on biological families that were composed of charismatic megafauna (particularly large mammals, monkeys, and birds). Plants were almost never held up as species that needed special conservation attention (although this is changing, with more recent attention being paid to especially diverse and endangered flora, such as conifers; see Nguyen Tien Hiep et al. 2004). This is in some ways a function of the funding constraints that conservation NGOs are subject to; large "cute critters" bring in more money in grants and donations than fungi and bugs, as several conservationists pointed out to me. The use of "cute" animals as stand-ins for biodiversity writ large is not confined to Vietnam, as many have noted (Humphries, Williams, and Vane-Wright 1995). The association of particular megafauna—tigers, lions, or rhinoceroses—with ideas of true "wilderness" and places removed from human influence is widespread (Mullin 1999). Many works have shown the predominance of attention to a few select categories of animals in international conservation efforts to the exclusion of other equally endangered life forms (Bonner 1993; Guha 1997). The reasons for this are varied and complex, but in part can be explained by the initial push for conservation coming from elite hunters in England and the United States (what we might call the "Teddy Roosevelt model" of conservation; Haraway 1985). Early conservation also was intimately tied up with colonial expansion in the tropics (MacKenzie 1988); Flora and Fauna International was originally founded in England in the early 20th century as the Society for the Protection of the Fauna of the Empire (Neumann 1996). Tropical visions of "Edenic" landscapes characterizes much early colonial efforts at conservation (Grove 1992, 1995), and these visions of paradise untouched by man can be found in much of the descriptions of places like Vu Quang today.

For international conservationists in Vietnam, the correlation be- tween "wild" animals and "natural" areas in need of conservation was very strong. Animals that were newly described by science such as the saola and giant muntjac were automatically assumed to be highly endangered; why else would they have eluded scientists for so long? Because these animals were presumed to be in great need of immediate conservation, wide swaths of landscapes they might potentially be found in began to be blocked off for "protection"; wild animals needed to be zoned off from humans and their encroachment on these wild areas. Similar attention has been paid by others to this rather arbitrary divide that supposes wildlife to be on the edges of the "social" and "society" (Whatmore and Thorne 1998). This divide is clearly reflected in conservation attention to species seen as "new" and "wild" (like the saola) rather than "old" and "domesticated" (like rice).

This is however a troubling and arbitrary partition, as the problem of loss of agrodiversity would seem to be one that conservation NGOs would have embraced. For example, attention to in situ agrodiversity has been increasing in many areas of the world (Brookfield and Stocking 1999) and among many conservation organizations; however, this type of attention has not been seen to date in Vietnam. This is partly because the main Western conservation NGOs working in Vietnam have been those with a history of working on primarily animal conservation: WWF, Flora and Fauna International, and BirdLife International. Other conservation groups that have been active globally on agriculture and land issues have not yet gained a foothold in the country, and local conservation groups are virtually nonexistent, with the exception of a few urban youth-oriented NGOs. There is a great need for organizations who can advocate for attention to more human-oriented environmental issues, like "brown issues" (pollution and urban problems) and agriculture issues, like genetic diversity of indigenous crops. There have been some small-scale projects, like the Community Biodiversity Development and Conservation Programme (CBDC), a modest project on genetic diversity of rice that is run out of an NGO in the Philippines (CBDC 2001). Interestingly, the few agrodiversity projects found in Vietnam have mostly been South–South ones, with NGOs and researchers in neighboring countries like Thailand, China, and the Philippines ex- changing research on this topic, while "Western" NGOs have not been involved.

This is unfortunate, as the preservation of indigenous species like floating rice, and the wild varieties that begat them, has several practical environmental implications of great importance. As modern rice varieties are continually improved, new genes often need to be obtained from wild relatives to add new traits, such as resistance to new pests. The loss of antique cultivars erodes the genetic diversity available to support breeding even more MVs. Although some have argued that MVs have not eroded the genetic diversity present in agriculture (Byerlee 1996), the overall consensus in much of the world is that genetic diversity of major crops is in fact in serious decline (Thrupp 2000). This seems to particularly be the case as marginal areas are converted to irrigated zones and planted to MVs of rice, as is the case in much of Southeast Asia, where local varieties are being replaced at a rapid rate (Morin et al. 2002).

Additionally, many studies have made a clear point that antique or traditional varieties are not "static"; they too undergo significant change over time as farmers experiment in their fields, and need to be continually cultivated, not simply preserved ex situ (Jackson 1995; Oldfield and Alcorn 1987). IRRI has conserved some varieties in ex situ gene banks in the Philippines, but recent research in the Mekong Delta indicates that ex situ preservation compared with in situ conservation produced significant differences in output—rice varieties kept at IRRI were very different from the same variety being actively managed in farmer trials (Tin, Berg, and Bjornstad 2001). As a recent study of agrodiversity in China notes, "Crop varieties are the result not only of natural factors, such as mutation and natural selection, but also, and particularly, of human selection and management ... farmers' decisions ultimately determine whether these populations are maintained or abandoned" (Zhu et al. 2003: 158).

We can contrast these dynamic and continually changing rice landraces and varieties that require human interaction with the discourse that surrounded the saola. Rather than being linked to human activity, the saola has been invoked as evidence of a lack of humans, as a symbol of "lost worlds" of untouched wilderness. The saola itself was described as biologically interesting because it was a "primitive taxa," a long-unchanging animal lost in the mists of time in Vietnam. Scientists described Vu Quang as a "biotically unique region where primitive taxa, long extinct elsewhere, have been able to survive into the late 20th century" (Groves et al. 1997: 335), a sentiment echoed by other popular articles in *Natural History* (Sterling, Hurley, and Bain 2003) and *The Sciences* (Drollette 2000) that have emphasized the "timelessness" of Vu Quang. This emphasis on the isolation and lack of human involvement with the saola (until the Western scientists found it) directly led to the assumption that to protect the animal, human involvement in the landscapes of saola habitat had to be prevented. This accounted for the emphasis on park boundaries and rangers in the VQCP, and the prohibition on human use of park resources once the reserve was demarcated (Kemp and Dilger 1997; Vu Van Dung et al. 1994).

Very few nature reserve management plans anywhere in Vietnam paid any attention to historical evidence of human use, assuming that "wild" animals and "wilderness" were coterminous, and Vu Quang was no exception. None of the descriptions of the "lost world" so heralded in media accounts ever mentioned that Vu Quang was once a timber reserve, that 20,000 people lived in and near the forests, that feeder roads to the heavily bombed Ho Chi Minh trail ran through these "pristine" forests, or that pitched battles were fought all along this border between the North Vietnamese Army and the U.S.-backed Lao Hmong guerilla army during the 1960s and into the 1970s (McElwee 2003). It is difficult to know if the press releases that declared Vu Quang to be a "lost world" were written in hyperbole to raise funds for conservation, or if the NGOs involved really truly did believe they had stumbled on an Edenic place untrammeled by man. Had the conservationists' association of "wild" animals, like the saola, with "wild places" led directly to these false imaginings about "lost worlds"?

Regardless of the reasons for the use of the "lost world" idea, in the end, the metaphor may have been its own undoing for true attempts to find solutions to protect Vu Quang's treasures. If Vu Quang was a lost world, there were no people in it, and if there were no people in it, there was no need to raise money for local economic development to make up for restriction of access to the park (as was the case in the first years of the VQCP). Given that very few people in the opening years of the VQCP paid any attention to the most dominant animal in the area—humans—it was no wonder that conservationists soon became disillusioned. Vu Quang was in fact no "lost world" but a forest landscape actively used by many people, and these people wanted a piece of the VQCP and continued access to the resources of the park. And so the metaphor that began the VQCP, that of a paradise on earth, had changed by the end of the VQCP, to that of a different type of lost area—a "paradise lost."

Conclusion

Real world decisions in the fields of environmental and agricultural management are directly affected by definitions and ideas of "domesticated" and "wild" species, as I have tried to show. Conservation

NGOs in Vietnam have defined their purview as protecting "wild" animals and "wilderness," some of which have been portrayed as mystic creatures unaffected by human changes for millennia. Now that these "pristine" areas are reputedly threatened by human actions there have been several projects with millions of dollars spent to protect them. Yet threats to species, varieties and landraces that have been clearly affected and manipulated by human intervention, like rice, have not been on conservation agendas in Vietnam.

Perhaps conservation-oriented NGOs and researchers have assumed that rice and agricultural diversity issues fall under someone else's purview, and are expecting IRRI, the Ministry of Agriculture, or other entities to advocate for biodiversity in agriculture as strongly as the conservation groups advocate for biodiversity in "wilderness." To some degree, these other institutions have indeed paid some attention to the issues, as evidenced by the conference in 2000 on in situ biodiversity conservation in crops that was held in Hanoi. But the assumption that agriculture is someone else business, not an issue for environmental NGOs, remains a seriously problematic assumption, given the evidence laid out in this chapter about the environmental effects of HYVs. It appears that the conservation organizations' epistemological divide between the "wild" and the "domestic" has similarly put blinders on their view of what is the "environment" and what is "agriculture"—the former being what they are primarily interested in. To be fair, these organizations have limited budgets and often small staff for such a large and populous country as Vietnam; these NGOs cannot possibly be expected to advocate for every single conservation cause. Thus, the main emphasis in conservation in Vietnam has been species conservation, and primarily conservation of endangered animal species, where the need has seemed most pressing.

Yet changes in agriculture have been happening extremely rapidly as well, and directly effect environmental conditions for many species, such as fish biodiversity that may be affected by the massive use of pesticides and herbicides that most MVs now require (Mai Dinh Yen 1997). Furthermore, should increased flooding be the norm in the Mekong, as many have predicted thanks to excessive deforestation in the upper reaches of the Mekong in Laos, Burma and China, how will the millions of farmers in the Mekong Delta be affected? The use of traditional varieties were uniquely suited to be able to adapt to these changes in the river levels. The new HYV are not. Studies in the Philippines noted that during several lean years following El Nino flooding, farmers had significantly reduced seed stocks, particularly of traditional varieties, because they were either forced to consume or sell all of their reserved stock owing to poverty, or the flooding conditions ruined their storage areas (Morin et al. 2002). Should cyclical flooding in Southeast Asia and elsewhere continue, as El Nino patterns suggest, and should trends in global warming research hold true that sea levels may rise as well, many areas of the Mekong Delta may experience water levels unsuitable for continued cultivation of HYV.[2] Should that occur, will there still be floating rice varieties that farmers can fall back on, or will they too have been wiped out by natural disasters combined with long-term trends toward MVs, as was the case in the Philippines noted by Morin et al. (2002)?

We can note other potential scenarios as well. Faced with loss of agricultural crops in previous years of floods in places like Central Vietnam, many farmers have turned to other activities such as fuelwood and charcoal collecting or hunting in protected forests to make ends meet. These activities will likely increase if large numbers of HYVs fail in floods. Agricultural changes thus have direct impacts on "natural" biodiversity in a myriad of ways, as people try to find ways to make up for losses of agricultural livelihoods.

Conservation organizations so far have not shown themselves to be interested in this particular challenge, calling into question their ideas of what "biodiversity" is and how it might be preserved. Has the attention to "wild" species, to the exclusion of the value of "domesticated" species, foreshadowed future problems for millions of farmers in the Mekong Delta? Will some NGOs, perhaps new, locally founded Vietnamese ones, recognize the links between agriculture and biodiversity, and thus step up to be the champions of the preservation of traditional varieties and agrodiversity as strongly as the older NGOs have championed the saola? Only time will tell.

Postscript

After this chapter was written, scientists with the IRRI and several California universities made an announcement in the August 2006 issue of *Nature* that they had identified the gene that regulates water tolerance in rice. (The entire genetic code of *O. sativa* spp, *japonica* was successfully mapped by IRRI researchers in 2005.) The new study identified the particular piece of rice DNA that affects flood tolerance in some specific local varieties. These were not the floating varieties I have described here that can raise their panicle as waters rise, but, rather, varieties that are "submergence-tolerant"; that is, they can survive with their panicle underwater for nearly 2 weeks and then continue to grow once the water subsides. One specific gene was found that creates this "submergence toleration" and that gene was manipulated and crossbred into other varieties that did not originally have it. The authors report that this new transgenic variety with flood tolerance, if widely used, "is expected to provide protection against damaging floods and increase crop security for farmers" (Xu et al. 2006).

Although this new genetically modified variety could potentially help avoid some of the disastrous scenarios of rising sea levels I have identified here, questions remain if this rice will indeed be used widely given that it is a type of genetically modified organism (GMO). It is possible that strong anti-GMO sentiment in Europe and elsewhere could hinder its use. Some international conservation NGOs, such as Greenpeace, have been very vocal about GMOs and oppose their use. Many types of GM rice are currently being developed and tested, such as the so-called "golden rice" that features gene manipulation to increase vitamin A delivery. However, no variety of GM rice is being grown anywhere in the world on a commercial basis yet (Lu and Snow 2005), in contrast with GM corn and soybeans, which are more widespread. It is very difficult to predict what the impact of these GMOs will have on the types of understandings of domestication that this volume has highlighted.

Notes

1. Hybrid rice is produced by crossing two inbred "parent" varieties, which yields 10–20 percent higher production over other HYVs in the first generation. Hybrid rice is not, however, genetically modified (GM) rice.

2. The impacts of global warming on crops such as rice are not well-known. Obviously sea level rises will affect some rice-growing areas, but other effects are less well studied. Rice is likely to become more productive if CO^2 rises, but not necessarily if temperatures do. And, of course, rice growing is also a contributor to global climate change in that methane, a potent greenhouse gas, is a significant by-product of rice paddies (Neue et al. 1995).

References

Berg, H. 2001. Pesticide use in rice and rice-fish farms in the Mekong Delta, Vietnam. *Crop Protection* 20 (10): 897–905.

Bonner, R. 1993. *At the hand of man: Peril and hope for Africa's wildlife.* New York: Vintage.

Briscoe, D. 1992. Signs of new mammal species are among finds in Vietnam's "lost world." *Washington Post,* July 28: A10.

Brookfield, H., and M. Stocking. 1999. Agrodiversity: Definition, description and design. *Global Environmental Change* 9: 77–80.

Byerlee, D. 1996. Modern varieties, productivity, and sustainability: Recent experience and emerging challenges. *World Development* 24: 697–718.

Cao Hung. 1999. Gan 300ha rung bi tan pha, ai chiu trach nghiem? (Nearly 300ha of forest are deforested, who has responsibility?). *Lao Dong* [Labor], Hanoi, August 10.

Community Biodiversity Development and Conservation Programme (CBDC). 2001. *Seed supply system in the Mekong Delta, Vietnam.* (CBDC) Programme—Vietnam Project. Technical Report No. 21. Quezon City, Philippines: Southeast Asia Regional Institute for Community Education.

Constable, J. 1982. Visit to Vietnam. *Oryx* 16: 249–254.

Cummings, R. C. 1978. Agricultural change in Vietnam's floating rice region. *Human Organization* 37: 235–245.

Daming, H. 1995. *Facilitating regional sustainable development through integrated multi-objective utilization, management of water resources in the Lancang-Mekong River Basin.* Kunming: Center for Environmental Evolution and Sustainable Development.

Dao The Tuan. 1985. Types of rice cultivation and its related civilizations in Vietnam. *East Asian Cultural Studies* 24: 41–56.

Dap, B. H. 1980. *Cay lua Vietnam* [Rice of Vietnam]. Hanoi: Nha Xuat Ban Khoa Hoc Ky Thuat [Science and Technology Publishing House].

David, C., and K. Otsuka. 1994. Introduction. In *Modern rice technology and income distribution in Asia,* edited by C. David and K. Otsuka, 3–10. Boulder, CO: Lynne Rienner.

DeKoninck, R. 1979. The integration of the peasantry: Examples from Malaysia and Indonesia. *Pacific Affairs* 52 (2): 265–293.

Denning, G. L., and Vo Tong Xuan, eds. 1995. *Vietnam and IRRI: A partnership in rice research.* Los Banos, Philippines: IRRI.

Dove, M. R., and D. M. Kammen. 1997. The epistemology of sustainable resource use: Managing forest products, swiddens, and high-yielding variety crops. *Human Organization* 56: 91–101.

Drollette, D. 1999. "The Last Frontier": In Vietnam, zoologists discover animals long hidden by rugged terrain and political strife. *Newsday,* New York, April 27.

——. 2000. Lost and found. *The Sciences* 40: 16–20.

Eve, R. 1998. Vu Quang Nature Reserve: A link in the Annamite Chain. Hanoi: WWF.

Evenson, R., and D. Gollin. 2003. Assessing the Impact of the Green Revolution, 1960 to 2000. *Science* 300 (May 2): 758–762.

Farmer, B. H. 1986. Perspectives on the "Green Revolution" in South Asia. *Modern Asian Studies* 20: 175–199.

Farmer, B. H., ed. 1977. *Green revolution? Technology and change in rice-growing areas of Tamil Nadu and Sri Lanka*. Boulder, CO: Westview Press.

Grove, R. 1992. Origins of Western environmentalism. *Scientific American* 267 (1): 22–27.

——. 1995. *Green imperialism: Colonial expansion, tropical island edens and the origins of environmentalism, 1600–1860*. Cambridge: Cambridge University Press.

Groves, C., G. Schaller, G. Amato, and K. Khamkhoun. 1997. Rediscovery of the wild pig *Sus bucculentus*. *Nature* 386 (6623): 335.

Guha, R. 1997. The authoritarian biologist and the arrogance of anti-humanism: Wildlife conservation in the Third World. *The Ecologist* 27 (1): 14–20.

Haraway, D. 1985. Teddy bear patriarchy: Taxidermy in the garden of Eden, New York City, 1908–1936. *Social Text* 11: 20–64.

Hardcastle, J., S. Cox, Nguyen Thi Dao, A. G. Johns. 2004. *Rediscovering the Saola: Proceedings of a workshop*. Nghe An: WWF Indochina, SFNC Project, Pu Mat National Park.

Hargrove, T. R. 2002. *A dragon lives forever: War and rice in Vietnam's Mekong delta 1969–1991, and beyond*. New York: Ivy Books.

Heong, K. L., M. M. Escalada, and V. O. Mai. 1994. An analysis of insecticide use in rice: Case studies in the Philippines and Vietnam. *International Journal of Pest Management* 40 (2): 173–178.

Hickey, G. 1964. *Village in Vietnam*. New Haven, CT: Yale University Press.

Hoang Tuyet Minh. 2000. Rice biodiversity in Vietnam—Current status and the causes of erosion. In *In-situ conservation of native landraces and their wild relatives, conference proceedings*, 58–60. Hanoi: Institute of Agricultural Genetics Vietnam.

Hossain, M., Duong Ngoc Thanh, and F. B. Gascon. 1995. Change from deepwater to irrigated rice ecosystem in the Mekong River Delta: Impact on productivity and on farmers' income. In *Vietnam and IRRI: A partnership in rice research*, edited by G. L. Denning and Vo Tong Xuan, 263–273. Los Banos, Philippines: IRRI.

Humphries, C. J., P. H. Williams, and R. I. Vane-Wright. 1995. Measuring biodiversity value for conservation. *Annual Review of Ecology and Systematics* 26: 93–111.

Institute of Agricultural Genetics Vietnam. 2000. *In-situ conservation of native landraces and their wild relatives, conference proceedings.* Hanoi: Institute of Agricultural Genetics Vietnam.

Jackson, M. 1995. Protecting the heritage of rice biodiversity *GeoJournal* 35: 267–274.

Jacobs, J. 1996. Adjusting to climate change in the Lower Mekong. *Global Environmental Change* 6: 7–22.

Jacoby, E. H. 1972. Effects of the "green revolution" in South and South-East Asia. *Modern Asian Studies* 6: 63–69.

Kemp, N., and M. Dilger. 1997. The saola *Pseudoryx nghetinhensis* in Vietnam—new information on distribution and habitat preferences and conservation needs. *Oryx* 31: 37–44.

Khush, G. 1995. Modern varieties—their real contribution to food supply and equity. *GeoJournal* 35: 275–284.

Le Huan. 1999. Vi sao rung Tay Nguyen bi pha [Why are the central highlands being deforested?]. *Lao Dong,* Hanoi, May 14: 1.

Linden, E. 1994. Ancient creatures in a lost world. *Time Magazine* 143 (25): 52–54.

Logan, W. 1971. How deep is the Green Revolution in South Vietnam? The story of the agricultural turn-around in South Vietnam. *Asian Survey* 11: 321–330.

Lu, B. R., and A. Snow. 2005. Gene flow from genetically modified rice and its environmental consequences. *BioScience* 55: 669–678.

MacKenzie, J. 1988. *The empire of nature: Hunting, conservation and British imperialism.* Manchester: Manchester University Press.

Mai Dinh Yen. 1997. Freshwater biodiversity and the measures for conservation in the Red River Delta of Vietnam. *Acta Hydrobiologica Sinica* 21 (Suppl.): 166–169.

McElwee, P. 2003. *"Lost worlds" or "lost causes"?: Biodiversity conservation, forest management, and rural life in Vietnam.* Ph.D. dissertation, Departments of Forestry and Environmental Studies and Anthropology, Yale University.

McElwee, P., and M. Horowitz. 1999. *Environment and society in the lower Mekong River basin: A landscaping review.* Binghamton, NY: Institute for Development Anthropology.

Milius, S. 1999. And now there are two striped rabbits. *Science News* 156 (8): 116–119.

Miller, Frank C. 1977. Knowledge and power: Anthropology, policy research, and the green revolution. *American Ethnologist* 4 (1): 190–198.

Morin, S. R., M. Calibo, M. Garcia-Belen, J. L. Pham, and F. Palis. 2002. Natural hazards and genetic diversity in rice. *Agriculture and Human Values* 19: 133–149.

Mullin, M. 1999. Mirrors and windows: Sociocultural studies of human–animal relationships. *Annual Review of Anthropology* 28: 201–224.

Neue, H. U., L. H. Ziska, R. B. Matthews, and D. Qiujie. 1995. Reducing global warming—the role of rice. *GeoJournal* 35: 351–362.

Neumann, R. 1996. Dukes, earls and ersatz edens: Aristocratic nature and preservationists in colonial Africa. *Environment and Planning* 14: 76–98.

Nguyen Huu Chiem. 1994. Former and present cropping patterns in the Mekong Delta. *Southeast Asian Studies (Japan)* 31: 345–384.

Nguyen Khac Vien. 1985. *Southern Vietnam 1975–1985*. Hanoi: Foreign Languages Publishing House.

Nguyen Thi Song An. 1996. *Agricultural reforms in Vietnam: The impact on rice production and farmers in the Mekong Delta (1976–1995)*. Ph.D. dissertation, the Institute of Social Studies, The Hague.

Nguyen Tien Hiep, Phan Ke Loc, Nguyen Duc To Luu, P. Thomas, A. Farjon, L. Averynov, and J. Regalado. 2004. Vietnam conifers: Conservation status review 2004. Hanoi: Fauna and Flora International, Vietnam Conservation Support Programme.

Oka, H. I. 1975. Floating rice, an ecotype adapted to deep-water paddies—A review from the viewpoint of breeding. In *Rice in Asia,* edited by Y. Ochi, 277–287. Tokyo: University of Tokyo Press.

Oldfield, M., and J. Alcorn. 1987. Conservation of traditional ecosystems. *BioScience* 37 (3): 199–208.

Pantulu, V. R. 1986. The Mekong River system. In *The ecology of river systems,* edited by B. Davies and K. Walker. Dordrecht, the Netherlands: DR W. Junk Publishers.

Pimentel, D., and M. Pimentel. 1990. Comment: Adverse environmental consequences of the green revolution. *Population and Development Review* 16: 329–332.

Pingali, P. L., Nguyen Tri Khiem, R. Gerpacio, and Vo Tong Xuan. 1997. Prospects for sustaining Vietnam's reacquired rice exporter status. *Food Policy* 22: 345–358.

Pingali, P. L., and Vo Tong Xuan. 1992. Vietnam: Decollectivization and rice productivity growth. *Economic Development and Cultural Change* 40: 697–718.

Rao, S. A., C. Bounphanousay, J. M. Schiller, A. P. Alcantara, and M. T. Jackson. 2002. Naming of traditional rice varieties by farmers in the Lao PDR. *Genetic Resources and Crop Evolution* 49: 83–88.

Rerkasem, B., and K. Rerkasem. 2005. *On-farm conservation of rice biodiversity.* Paper presented at the FAO international workshop "In-situ conservation of native landraces and their wild relatives," Bangkok, August 29–September 2.

Schaller, G. 1998. On the trail of new species. *International Wildlife* 28 (4): 36–44.

Schaller, G., Nguyen Xuan Dang, Le Dinh Thuy, and Vo Thanh Son. 1990. Javan rhinoceros in Vietnam. *Oryx* 24 (2): 77–80.

Singh, R. B. 2000. Environmental consequences of agricultural development: A case study from the Green Revolution state of Haryana, India. *Agriculture, Ecosystems and Environment* 82: 97–103.

Socialist Republic of Vietnam. 1995. *Biodiversity action plan for Vietnam.* Hanoi: Socialist Republic of Vietnam and Global Environment Facility.

Sterling, E. J., M. M. Hurley, and R. Bain. 2003. Vietnam's secret life. *Natural History,* March: 50–59.

Tangley, L. 1997. A new brief for nature: Science is revising the politics of biodiversity. *U.S. News and World Report,* October 10: 68.

Terborgh, J. 1999. *Requiem for nature.* Washington, DC: Island Press.

Thrupp, L. A. 2000. Linking agricultural biodiversity and food security: the valuable role of sustainable agriculture. *International Affairs* 76: 265–281.

Time Magazine. 1992. Journey into Vietnam's lost world. *Time Magazine* 140 (6): 22–27.

Tin, H. Q., T. Berg, and A. Bjornstad. 2001. Diversity and adaptation in rice varieties under static (ex situ) and dynamic (in situ) management: A case study in the Mekong Delta, Vietnam. *Euphytica* 122: 491–502.

Tran Thi Ut. 2002. *The impact of the green revolution on rice production in Vietnam.* Paper presented to Foundation for Advanced Studies on International Development workshop "Green revolution in Asia and its transferability to Africa," Tokyo, December 8–10.

Tran Thi Ut and K. Kajisa. 2006. The impact of green revolution on rice production in Vietnam. *The Developing Economies* 44 (June): 167–189.

U.S. Engineer Agency for Resources Inventories. 1968. *Atlas of physical, economic, and social resources of the lower Mekong basin.* New York: U.S. Engineer Agency for Resources Inventories, Tennessee Valley Authority, and USAID's Bureau for East Asia, Committee for Coordination of Investigations of the Lower Mekong Basin.

Vo Tong Xuan. 1995. History of Vietnam-IRRI cooperation. In *Vietnam and IRRI: A partnership in rice research,* edited by G. L. Denning and Vo Tong Xuan, 21–29. Los Banos, Philippines: IRRI.

Vo Tong Xuan, Le Thanh Duong, Nguyen Ngoc De, Bui Chi Buu, Pham Thi Phan, and D. W. Puckridge. 1995. Deepwater rice research in the Mekong River Delta. In *Vietnam and IRRI: A partnership in rice research,* edited by G. L. Denning and Vo Tong Xuan, 179–190. Los Banos, Philippines: IRRI.

Vo Tong Xuan and S. Matsui, eds. 1998. *Development of farming systems in the Mekong delta, Vietnam.* Ho Chi Minh City: Ho Chi Minh City Publishing House.

Volker, A., ed. 1993. *Hydrology and water management of deltaic areas.* Rotterdam: A. A. Balkema.

Vu Van Dung, Pham Mong Giao, Nguyen Ngoc Chinh, Do Tuoc, and J. Mackinnon. 1994. Discovery and conservation of the Vu Quang Ox in Vietnam. *Oryx* 28: 16–21.

Wege, D. C., A. J. Long, and Mai Ky Vinh. 1999. *Expanding the protected areas network in Vietnam for the 21st century: An analysis of the current system with recommendations for equitable expansion.* Hanoi: BirdLife International.

Whatmore, S., and L. Thorne. 1998. Wild(er)ness: Reconfiguring the geographies of wildlife. *Transactions of the Institute of British Geographers* 23 (4): 435–454.

World Bank. 1994. *Cambodia: From rehabilitation to reconstruction.* Washington, DC: World Bank Country Operations Report.

Xu, K., X. Xu, T. Fukao, P. Canlas, R. Maghirang-Rodriguez, S. Heuer, A. Ismail, J. Bailey-Serres, P. Ronald, and D. Mackill. 2006. Sub1A is a ethylene-response-factor-like gene that confers submergence to tolerance to rice. *Nature* 442 (August 10): 705–708.

Yapa, L. 1993. What are improved seeds? An epistemology of the Green Revolution. *Economic Geography* 69: 254–273.

Zhu, Y., Y. Wang, H. Chen, and B. R Lu. 2003. Conserving traditional rice varieties through management for crop diversity. *BioScience* 53: 158–162.

Feeding the Animals

Molly H. Mullin

In 2003, Mars Inc. launched its "Inner Beast" campaign of print and television ads for its "Whiskas" brand of cat food in North America. Featuring digitally manipulated images of a cat hunting a buffalo in one version, a zebra and a herd of impalas in others, the 30-second television commercials and two-page print ads whimsically juxtaposed wild and domesticated, human and animal, common and extraordinary. Cats were domesticated only relatively recently, explained the ads, and cats' dietary needs must be understood in relation to that recent domestication and to millions of years of evolution preceding it. Citing 30 years of scientific research, Whiskas promised to calibrate the gap between housecat and lion in a manner pleasing to both cat and human. With images reminiscent of wildlife documentaries, safaris, and zoologists at work, the ads connected caring for pets with knowing and revering the wild.

The Inner Beast ads can be considered a statement in an ongoing discussion concerning the relationship between wild and domesticated animals. One might imagine this relationship of diminishing importance among contemporary humans, especially among those living in contexts in which it is easy to see everything as domesticated and where the wild would seem to matter little in everyday life. The wild however, serves as a powerful resource in commercial and popular culture and it is not only anthropologists and other scholars and scientists who are concerned with understanding (and reconsidering) domestication and its consequences.

Anthropology has a long history of attending to how people understand and make use of the wild. However, the existing literature can be of limited assistance when aiming for ethnography that is accommodating of change, variation, and a multiplicity of meanings and interpretations. For example, following Lévi-Strauss, anthropologists have observed that domestic animals occupy an ambiguous status among humans (Leach 1964; Lévi-Strauss 1971: 468). This is said to be especially true of dogs, described by some anthropologists as living at the threshold of culture and nature (Vidas 2002; White 1991: 12). Increasingly, however, the status of so many animals appears uncertain or in dispute that it has become difficult to accept that the status of anything was ever unambiguous. Here, I explore particular versions of the relationship between wild and domesticated animals and how some people have come to see this relationship as problematic and contested. There are certainly long-standing patterns, and some emerging trends, in how the wild and the domesticated are imagined. But people continue to fashion new meanings and uses for these categories in ways that defy expectations.

Donna Haraway has argued that regardless of scientific validity, narratives of domestication or "co-evolution" function as powerful "origin stories" with more at stake in competing versions than might be readily apparent. At stake, she notes, is the relationship of nature and culture and "who and what gets to count as an actor" (Haraway 2003: 27). But who, apart from scholars and the dog breeders that Haraway considers, traffics in these stories and why? One way that "domestication stories" are being used is to legitimate consumer choices. In a context in which there is great emphasis on individual choice, choice informed by expert authorities and assessments of risk, domestication stories offer the promise of assessment based not, or not only, on corporate sponsored research in laboratories but on knowledge seemingly less shaped by corporate interests. If, among consumers, the science of the laboratory has lost some of the authority it wielded a decade ago, the science of the field (including that of archaeology and zoology) has gained appeal. One can see this shift in the recent history of thinking about and marketing pet food.

Contemporary pet keeping, when it is not being celebrated, is often caricatured as a matter of rampant anthropomorphism and conspicuous consumption. These abound, to be sure, but the ways in which people imagine even "companion animals'" relationship to the wild is more complex than captured by the caricature (Grier 2006: 182–185). For some, the "wild" can be a means of reacting against anthropomorphism and consumerism. I do not argue that it is an effective means of doing so, but I do try to show some of the ways the wild has become important for people negotiating relationships with other humans and with other species, particularly companion animal species, in a context of expanding consumer capitalism.

Domestication has been widely recognized by anthropologists as an important part of some of the most profound transformations, not only in humans' relationships with other species, but also in humans' relationships with one another. At times heralded as the hallmark of civilization and even of humanity, domestication has been credited with making possible new kinds of wealth, inequality, divisions of labor, and powerful conceptual resources with which to naturalize and legitimate patterns of differentiation and exploitation (Anderson 1998; Milliet 2002; Russell 2002; Thomas 1983). Especially when it is perceived as endangered, as it has so often in the past century, wildness has gained more positive connotations, as something worthy of protection and reverence. Although domestication is often perceived as superior to the wild, it has also been associated with servitude and degeneracy (Sax 2000, Thomas 1983: 44–45, Tuan 1984: 99). Whatever the particular attributes or how these terms are being used, "domestication" and the "wild" are capable of triggering complex and intense emotional responses.

There are a number of reasons why this complexity and intensity may be amplified in considerations of animal feeding. Eating is something that humans quite obviously share with animals. Such commonality has been threatening in contexts with a heavily patrolled human–animal boundary. Norms pertaining to food and eating have often been used to mark the boundaries around what is acceptably human. Food links humans to other beings, human and animal; it creates moral obligations, it is rich with meaning and emotional power (Knight 2005: 232). The children's story that inspired the title of this volume provides an example of some of the power attached to food and feeding: unruly Max goes off to "where the wild things are" after threatening to eat his mother and being sent to bed with no dinner. It is the smell of "good things to eat" that lures him home—and there he finds his supper, "still hot," awaiting him (Sendak 1963).

For Max, enjoying "good things to eat" requires proper behavior. The desire for this enjoyment links him to others on whom he depends as providers. In terms of how people think about feeding animals, it is important that it is sometimes possible to tame wild animals by feeding them or at least to change their behavior in significant ways, thus encouraging connections between domestication and control, wildness and freedom. I am not supporting the notion that taming is equivalent to domestication (Russell [this volume] reviews reasons why it is not), but there are good reasons why people associate the two, helping to make the feeding of animals a morally charged practice. Also, as John Knight (2005) notes in his study of monkey feeding at parks in Japan, feeding animals provides a way for people to experience a rare kind of intimacy with animals, an intimacy that sometimes can turn out to be more complicated and less appealing than expected. Although feeding may hold the promise of intimacy and the possibility of control, people remain less aware of the extent to which it is not just animals that are affected by feeding, that feeding practices affect the feeders as well as the fed.

There is also the matter of what is being fed and eaten: The most popular companion animals have included dogs and cats, who can eat or be fed the bodies of other animals—a practice that in many parts of the world is quite strongly associated with the wild, and even among meat eaters as something at least slightly alien to the civilized and tame, and a practice capable of eliciting a combination of horror, disgust, fascination, and admiration. The eating habits of carnivores evoke even more intense responses among consumers living in contexts in which "butcher shops" have almost become a thing of the past and meat is rarely seen on the bone or in any shape that suggests it came from an animal. In such contexts, the crunching of bones and flesh is most familiar on the soundtracks of wildlife documentaries or horror films.[1]

Finally, in the period of the pet food industry's most rapid expansion, the feeding of pets and the buying of pet food became part of a gendered division of household labor (Grier 2006: 84). If for men, domestication and domesticity united in commercial ideals of the lawn (Wigley 1999), for women, they came together in pet food. Pet food was marketed to women alongside infant formula, baby food, cosmetics, and household cleaning products. Pets became family members (at least in theory) and like the rest of the family, taking care of them required pleasing them. Just as the social and political order surrounding such messages has been contested, so also have been messages about domesticity, domestication, and the feeding of animals (see fig. 11.1).

Feeding and the Laboratory

In the second half of the 20th century, many people with no prior experience of caring for animals were acquiring pets. According to Jackson Lears, "During and after World War II, as American business set out to recapture much of the prestige it had lost during the Depression,

Figure 11.1. *Good Housekeeping,* June 1964. "Come, Home Tom," was part of a series of print and television ads featuring a woman trying to get her cat to come home.

corporations reclaimed their role as benevolent mediator between science and the suburban family" (1994: 124). The growing pet food industry, originally the product of the grain industry and meat packing industry in late 19th-century Britain and North America (Grier 2006: 284–285), encouraged pet owners to believe their animals needed special, species-specific food, best formulated by scientists in laboratories. Cat food, whether formulated in a lab or anywhere else, was slower to be developed and to catch on than dog food. Even with dog food, there have always been holdouts, especially among rural people and among breeders. Both groups are likely to have experience with older feeding methods and to recognize authorities other than product manufacturers and advertisers.

A television program that aired in the United States in 2001 illustrates the view that the feeding of pets should be a matter of laboratory science. It also gives a sense of how this view has lost some of its power. Public Television opened its new season of the series "Scientific American

Frontiers" that year with an episode on how "pets are being helped by technology and vice versa" (*TV Guide* 2001: 139). After segments on computer software used to teach people how to train herding dogs, how dogs' noses detect bombs, and the potential of video to relieve the boredom of caged parrots, the episode concluded with a visit to Hill's Pet Nutrition, Inc., known for its Science Diet and Prescription Diet brands of pet foods (not mentioned in the episode, Hill's has been owned by Colgate Palmolive since 1976, a fact that would surprise many consumers, because of Hill's association with veterinary clinics and its opposition, in consumers' minds, with "commercial" pet food). This segment of the show, entitled "The Bite Stuff," focused on the company's painstaking efforts to develop a dog food, said to be available "by prescription" from a veterinarian, and demonstrated to reduce dental plaque and bad breath. Viewers heard Hill's Director of Research Dan Richardson explain why periodontal disease is more rule than exception in dogs. Dogs' ancestors, he said, lived on a diet of small animals, a diet similar, he noted with a nod toward surrounding fields, to that of contemporary coyotes. Viewers were shown how to address the problems ensuing from domestic dogs' diet of canned meat and dry kibble with a toothbrush— but it was made clear what an onerous chore it is to brush even the teeth of compliant test-beagle Wendy. Hence, the Hills' company's drive to produce the product finally emerging from multiple feeding trials, taste tests, and sniffing-of-the-test-beagles' breath tests (performed by a team of women in white lab coats). Domestication creates problems, it was suggested, but technological innovation can solve these. We are meant, not just to be impressed, but also amused—amused by the anthropomorphism (Wendy, etc.) as well as by the spectacle of scientific know-how being applied to problems that, because they have to do with pets, would strike many viewers as at least a little silly.

In pet food company self-presentations, ironic humor is a relatively new development. The theme of food as the product of science, and domestication as something of a problem to be solved, goes back farther. In the early 20th century, when pet food companies had yet to produce "complete" foods for dogs or cats, references to science were common but very basic. Advertising for the Thoroughbred Dog Biscuit—intended, like rival brands of the late 19th and early 20th century, to be combined with food left to the discretion of dogs' owners—declared the product to be "the most scientific prepared food for a dog, prepared after twenty years of study and practice with these animals, and if fed as advised, will undoubtedly keep an animal healthy for a long life" (Thoroughbred Dog Biscuit n.d.).

By the 1930s, companies were including more detail. "DOGS AND SCIENCE DECIDES WHAT'S BEST FOR YOUR DOG!" announced Ralston Purina. This statement accompanied a photograph of the "Purina Experimental Kennels" in Missouri, described as "part of the world's largest dog feeding experiment to find out just what food ingredients really are best for your dog. The dogs show us which food ingredients they like best, which keep them in the finest condition, produce the best coats, etc." (Ralston Purina Company 1937: 35). Similar to more recent marketing efforts, the Purina book highlighted one ingredient making their brand superior (vitamin A, or "Pur-a-tene" [1937: 33]). During the same period, Gaines advertised, with pictures of leaping dogs, food containing "Viactron." Although made-up names for ingredients have become less common in the pet and human food business, "key ingredient" marketing survives. More recent examples have been Omega-3 fatty acids and other "nutraceuticals" such as glucosamine and the shells of green-lipped mussels, claimed to benefit animals with arthritis.

Although the theme of the Inner Beast ads, that domestication requires science to solve its problems, is not new, earlier versions placed more emphasis on the laboratory than the field as the source of solutions. Consider a publication produced by a Mars-owned company, in the late 1950s, on "Cats, Dogs, and Diets." In keeping with a trend a historian of advertising has described as "imperial primitivism" (Lears 1994: 146–147), a chapter on "The Modern Approach," reports that cats and dogs,

> in spite of thousands of years of becoming and being domesticated, still live to some extent an "unnatural" life. They must rely on others for the provision of their food, for instance, instead of feeding themselves at nature's dictation. It is no accident that the ownerless pariah dogs of the Orient, living much nearer the state of nature, are incomparably freer from disease than Western pets (Chappie, Ltd. n.d.: 9).

The publication says no more about the pariah dogs, but describes the "modern scientific approach" in the Chappie Research Laboratory, in which men in white coats formulate diets containing the right "nutritive components" in ways that appeal to dog and cat palates. The white clothing, it is explained, is worn not as "a mere gesture to impress visiting parties (which it never fails to do), but as a practical aid to maintaining corporate works hygiene" (Chappie, Ltd. n.d.: 11). The emphasis on hygiene, whiteness, machines, and assembly lines is more pronounced than one would find in more recent advertising, although white lab coats still abound, especially in ads for brands like Science Diet (some newer brands forego the medical approach and instead feature pictures of food that people would eat—carrots, peas, wheat, steak). Also dating the Chappie publication was a chapter extolling the advantages of whale meat ("whales are remarkably free from disease; tuberculosis, for instance, is unknown among them"), with proud mention that Chappie processes into dog food half of Britain's annual import of whale meat (n.d.: 15). U.S. and British product labels from the 1940s and 1950s advertised not just whale meat, but also horse and rabbit meat, a practice that would end when horses and rabbits began being treated more like pets or companion animals and whales became part of the endangered wild.[2]

Human and Animal at the Petfood Forum

"Prescription" or "veterinary" diets are an example of a more general trend toward what is described within the pet food industry as "premium" and "super-premium" varieties. "Innovation drives growth," a consultant told her audience at the 2003 Petfood Forum in Chicago, an important annual event for the industry. Speaking on "The Future of the Global Pet Food Industry," the consultant identified "super premium" dog and cat food as the most rapidly expanding segment of a $46 billion industry. Like others at the meeting, she described the industry—with remarkable confidence given the state of the U.S. economy, some 10 days into the U.S.-led war in Iraq—as "recession proof." She attributed the growth in premium brands not to strong economies, but to a shift toward "pet humanization" and a rise in single-person households and low population growth (Crossley 2003). Throughout the proceedings at the forum, participants seemed to share Crossley's interest in putting a finger on changes occurring in people's relationships with animals and in identifying the significance for the pet food industry. This is not surprising given that the industry's very name reflects assumptions about human–animal relationships that were the product of a particular moment in history and have increasingly come into question.

At the time I attended the Petfood Forum, I had been doing a lot of research among what you might call companion animal food activists, people working to promote alternatives to commercial pet food. I was making an initial foray into what many of those activists would see as enemy territory. So I was somewhat surprised to find the Petfood

Forum a friendly gathering of hardworking people. This was not the sort of conference where the attendees are off at the golf course leaving lecture halls empty. The halls were full of people taking notes and asking questions, wearing nametags with first names in large letters, corporate affiliations in smaller print. Most were from the United States, but I also met industry people who had flown in from Korea, Colombia, and South Africa. The majority of participants were not working for pet food companies directly but in related industries—packaging, animal by-products, processing equipment.

At breakfast, I mentioned the topic of this chapter to a veterinarian developing a specialty in veterinary nutrition. She provides recipes for homemade pet food through her veterinary clinic in New England and spoke enthusiastically about her work with veterinary nutritionists who advertise their services to clients on the Internet. She offered me a title, for whatever I might write concerning the anthropology of pet food: "We used to eat'em, now we feed'em!" At another gathering, a chemist, standing out not just as a woman but also as an Asian American in a large crowd of white men, spoke to me of her 20 years of work for a major company. She talked about her research colony of 400 cats, her attempts to make food they would eat with recognizable ingredients. People do not want to serve their cats ground food any more, she explained with a giggle. Nowadays they want it to look like food they might eat themselves. Unlike many others I spoke to, she described people who buy pet food sympathetically and as if she imagined them as fairly similar to herself. Several of the industry people I met spoke of pet food buyers as slightly strange and part of a trend they found vaguely unsettling: "I love our dog, *but...*"

Some of those most eager to talk about changing relationships between humans and animals were in the packaging business. One explained that his business is all about tracking trends: the sudden rise in popularity of stand-up packages as an indicator of higher quality, and of pull-off tops on cans because people do not want to use can openers on pet food that they would then use on food for themselves. Another packaging man told me that if I really wanted to get a sense of changing notions about humans and animals I must attend the Holistic Veterinary Association meetings. It is almost all women, he said, a lot of lesbians, and they emphasize emotional connections between people and animals. He spoke with what seemed fond amusement about a client who makes a "super-premium" line of dog food. She would never come to this event, he said, she would see only people trying to sell the cheapest product they can for the largest profit.

I heard only faint hints of derision or frustration as industry people talked about trying to stay on top of changing attitudes, a frustration similar to what they expressed about keeping up with aspects of globalization—involving, for example, efforts to make packaging and labeling conform to a wider variety of ever-changing government standards. Change, I gathered, was nothing new to the more experienced. Before the focus on globalization (with "biosecurity" emerging as a prominent secondary concern), consolidation was the hot topic, back when the current "top five" (Mars, Nestlé, Procter & Gamble, Colgate Palmolive, and Del Monte) were buying up the hundreds of companies doing business from the 1930s to the 1960s.[3]

I was surprised not to hear more frustration regarding the changing status of the "pet," especially because the direction of change was more complex than one might imagine. There was, of course, much that seemed clearly in keeping with the well-known "pets-as-family-members" trend. There was, for example, in addition to the popularity of food that looked like meat and vegetable stew, the concern for "human-grade ingredients," as well as considerable confusion and disagreement about how those might be defined. The owner of "Three Dog Bakery" stood in front of his audience eating from a bowl with a spoon and announced that he was eating a chicken-and-rice dog food. If the audience did not find this amusing, it is perhaps because the man-eating-dog-food act is an old one in the business. He made the point that his customers expect their toddler to be able to snack on the dog's food without getting sick. Although suggesting that people expect dog and human food to be more similar than in the past, he did not specify the impetus for change (i.e., he did not say whether there is greater similarity perceived between humans and dogs or whether people are more aware of the risk of disease transmission, or some combination).

There were two Petfood Forum participants with perspectives that contrasted starkly with the majority. An employee for People for the Ethical Treatment of Animals (PETA), who told me this was the first time her organization had sent a representative to the event, raised questions about the cruelty of feeding trials. In an attempt to deflect such criticism from animal rights organizations, increasingly pet food companies pay other companies to conduct feeding trials. The 1937 Purina booklet that proudly featured photographs of research kennels was very much the product of another era.[4] The PETA representative appeared to be cordially received by other participants, but it is possible that she felt less so the following year when the event brochure listed a session on "Dealing with Activists and Fanatics."

Another participant of particular interest to me and who shared the PETA representative's outsider status, was Australian veterinarian Ian Billinghurst. Since self-publishing his book *Give Your Dog A Bone* in 1993 and selling it largely over the Internet, Billinghurst has become well-known among North American dog breeders as one of a number of authors advocating alternatives to commercial pet food. At the Petfood Forum, Billinghurst explained to his audience that he based his methods of feeding on "evolutionary principles" as well as on his experience growing up poor in Australia and feeding butcher scraps to the family pets. Mill owners wanting to sell more grain in the 19th century, Billinghurst argued, started the pet food business. But does it make sense, he asked, to feed a grain-based (corn, wheat, or rice) diet to what essentially are carnivores? He went on to advocate that dogs be fed "the basic diet of the wolf," including raw meat and bones, and cats a diet even more strictly carnivorous.

The PETA representative told me she found his thinking refreshing and especially liked the idea that diets for pets could be based not on endless feeding trials, but on knowledge of evolution and on observation of the eating habits of wild and feral animals. Industry people tended to be less impressed, a common objection being a lack of scientific evidence in the form of controlled random clinical trials. There was also some concern that Billinghurst's approach was not likely to be financially profitable for large corporations. With relatively little processing involved, versions of this feeding method are similar to what the industry is trying to combat in parts of the world where commercial pet food has yet to gain wide acceptance. Also, as Billinghurst readily acknowledged, in North America it has been easy for a large cottage industry to meet rising demand. In fact, Billinghurst was there hoping to interest one of the top five companies in purchasing his own fledgling raw pet food business, an enterprise that has faced stiff competition from small-scale poultry and rabbit farmers, butcher shops, and a network of women distributing raw meat and bones from freezers in garages. Only a major corporation, Billinghurst told me, would have the legitimacy to gain a much larger share of the market.

Among veterinarians, the raw and homemade pet food trend has been tremendously controversial and at times bitterly fought, at least in North America (I take no position on it, or any other feeding method, here or elsewhere). Concerns include the risk to animals and humans of bacterial and protozoal "contamination," digestibility, and whether such diets are nutritionally "balanced" (Mieszkowski 2006). I have talked to vets with no objection to raw food diets. One rural vet attributed her own relatively tolerant attitude to the fact that she could not stop her dog from eating rabbits caught in a nearby field. I have also met vets who find the trend infuriating (along with acupuncture, herbal remedies, and pet psychics). On a mailing list devoted to criticism of "alternative" veterinary medicine, from 2001 to 2003 raw feeding was the most divisive of all topics, with arguments every few months going on for days at a time. In addition to the concerns mentioned above, debates have included discussions of anticorporatism, standards of evidence, and fundamental beliefs about progress, nature, domestication, and evolution (e.g., Devaney 2003; Ramey 2003).

After following the raw and homemade feeding controversy for several years, I was a little surprised the audience at the PetFood Forum was not more derisive of Billinghurst and his ideas. Although pet food companies do occasional research on wild and feral animals' eating habits, at the Petfood Forum, the wild appeared a curiosity as well as a category with a remote possibility of proving to be profitable. Although skeptical about Billinghurst's claim of benefits to animal health as well as the potential for profit, some were intrigued, including an elderly and apparently very well-respected professor of animal science who claimed to have devised, after World War II, some of the industry's most important inventions. "So you really don't see problems with salmonella and E coli?" he asked Billinghurst after the presentation, in a jovial "well, I'll be darned!" tone. A young man from one of the industry leaders politely chided Billinghurst on his lack of evidence in the form of random, controlled clinical trials.

In suggesting that in the interests of selling grain, pet food companies had obscured relationships between companion animals and their wild kin, Billinghurst came closer than anyone else I encountered at the Petfood Forum to exploring how the industry has helped to shape people's understandings of the animals with whom they share their lives and households. One can, however, see many ways in which the industry has actively sought to shape sensibilities as well as keep track of them. "A boy needs a dog" and "a dog needs a boy," claimed a poem included in the 1937 Purina care and training manual, to offer just one example (Ralston Purina Company 1937: 4).

I was initially struck by the extent to which pet food industry people shared my interest in understanding shifts in human–animal relationships. Steven Kemper has described advertising executives he studied in Sri Lanka as "folk ethnographers" and as "professional observers who make a living by convincing clients that they understand how the natives think" (2001: 4). After spending so much time hearing from particularly outspoken opinionated veterinarians and anticorporate activists, I felt at home among people more interested in figuring out what people want and how they think than in telling them what to do. But their interest is limited in important ways. Perhaps it should go without saying that industry people would not be especially inclined to look critically at how the industry directly and indirectly has helped to shape people's sensibilities. But the limited interest also follows from the hasty way knowledge is produced and exchanged in an industry context. At the Petfood Forum and in the pet food industry literature, there is a lot of traffic in labels to describe changes in people's relationships with pets or companion animals. Analysts get paid to categorize the changes going on worldwide and in particular regions. "Humanization" is currently a popular label for shifts that should help facilitate industry expansion (Kvamme 2006: 6). But is it possible that what it means to be human might change along with perceptions of nonhuman animals and the technology that surrounds their feeding? Is the direction of change always so orderly? After developing more intense emotional bonds with animals and therefore wanting to take care of them as well as possible, might some people begin to rethink the significance of the fact that their companions belong to another species, a species with a very different evolutionary history from their own, perhaps with a different relationship to the "wild"?

Wild Feeding

When using the categories of the wild and the domesticated, people pick and choose elements from existing patterns in ways that make sense to them. They focus on what appear, from their perspective, to be areas of the relationship between wild and domesticated that deserve particular attention, ignoring aspects that others might find more vexing. Commercial advertising is certainly not the only influence over how they do this. I first encountered the trend Billinghurst is associated with—that is, a method of feeding dogs and cats modeled, at least in part, on wild and feral animals' eating habits—while researching dog breeding, particularly the development of dog breeds that are considered by their proponents to be wilder, less thoroughly domesticated, than more established breeds (in fact, breeds formed from those "pariah dogs" mentioned in that early Chappie literature). New dog and cat breeds proliferated at a rapid pace in the late 20th century, a proliferation accelerated by the Internet, with its capacity to unite far-flung groups of people with common interests. Even before the Internet, breed clubs had a tendency to split into factions, divided by tastes for particular traits (color, size, personality, and abilities). Factions have been occasionally successful in establishing new breeds. There are also breeders looking for new market niches and alternative spheres of influence. One trend is to develop new breeds by crossing established ones. This is a disputed practice, but enjoys some popular appeal. "Labradoodles" and "Goldendoodles," for example, supposedly combine the poodle's best traits with those of Golden or Labrador Retrievers (there is a similar phenomenon among cat breeders). Retrievers are among the breeds created, at least in a formal way, in Europe during the mid–19th century—a period when most dog breeds were designed for some sort of hunting or guarding, purposes quite different from the reasons that often get people interested in keeping dogs today. Although a minority, there continue to be people using dogs for hunting, herding, and flock guarding, and some of them, as Donna Haraway observes, "emphasize the importance to dogs of jobs that leave them less vulnerable to consumerist whims" (2003: 38).

In addition to breeds being formed from crossing older ones and from established breeds splitting, there are also dog breeds that have been formed from feral canine populations and given an identity associated with a region or nationality: the New Guinea Singing Dog, American Dingo, Thai Ridgeback, and the Canaan Dog. I have been especially interested in the Canaan Dog, not because of any fondness for the breed, but as a case study in processes of commodification and how people think about the wild and about domestication.

Several years before I attended the Petfood Forum, I interviewed a Canaan Dog breeder in an immaculate Milwaukee suburb. She belonged to a faction of Canaan Dog people who tended to play down the breed's "wildness" and to emphasize characteristics that another faction disparaged as those of "yet another American household pet" (trainability, medium size, etc.). Like many in both factions, she likened the breed to an endangered species and spoke of her commitment to "improving the breed" as a sort of public service. She mentioned, in passing, that she did not feed her dogs commercial food, but fed them according to a method she had learned about on the Internet, modeled after what wild dogs would eat, consisting largely of raw poultry parts. At the time, I wondered whether this trend, if it was a trend, was a way for Canaan Dog people to emphasize the uniqueness of the breed, the perception of uniqueness clearly being part of the breed's appeal. In keeping with the endangered species analogy and the notion of the breed as less domesticated than others, it made sense that fanciers might

see the animals as so different from other dogs that they needed to be
fed a completely different kind of food.

I learned however that the breeder in Milwaukee was part of a trend
not confined to, or especially popular at the time, among fanciers of
the "primitive" dog breeds. The majority of Canaan Dog people that I
met between 1998 and 2002 were happily feeding various commercial
brands, more than one breeder at dog shows explaining that they fed
one "super-premium" brand to those currently on the show circuit or
breeding, and another, less expensive brand to the rest. Many Canaan
Dog people I spoke with were able to imagine the project of breeding—
with its pedigrees, judges, ribbons, contracts, and endless arguments
about official standards—as a way of preserving wildness, but it was
not lost on them that this might appear contradictory and problematic
and they were willing to discuss at length their positions on how things
should best proceed. Few at the time were devoting the same attention
to thinking about their method of feeding and its relationship to the
wild.

I quickly discovered that many other people in dogdom were doing
so. Cat people too. I have encountered these wild feeders on electronic
mailing lists; websites; in veterinary clinics; at a seminar I attended
in Canada taught by Ian Billinghurst and another seminar in Atlanta
offered by another Australian vet, Tom Lonsdale; and in interviews.[5]
They have mostly been women, just as, since the early 1970s, the majority of North Americans involved in dog breeding, animal rescue, and
other companion animal work have been women. Otherwise, their
diversity has been striking. Mostly from Australia and North America, but
including some Europeans and occasionally people from other regions,
they include breeders, agility competitors, hunters, farmers, vegetarians,
blind people feeding service dogs, and occasional PETA members. They
seem to share appreciation for emotional connections with individual
animals and see themselves as devoted to their well-being in a much
more overt way than you are likely to see among people keeping animals
for more utilitarian purposes (the same attitude that was spoken of with
some amusement by participants at the Petfood Forum). Otherwise,
however, I have found them to have such different relationships with
animals that I have been intrigued to find them willing to overlook their
differences, finding camaraderie in their marginality (at least in North
America), only to become divided, sometimes bitterly, over countless
other issues, particularly perceptions of risk (e.g., from zoonotic diseases)
and the specifics of how knowledge of wild carnivores and their eating
habits should be applied to dogs and cats.

Alternatives to commercial food have risen in popularity at the same time as premium and ultrapremium brands. Pet food companies, like those selling food for humans, have encouraged people to pay close attention to choices regarding food, and to see food as closely tied to health and well-being. The choices have also greatly multiplied, at least in areas where the pet food industry has a strong foothold. It is possible that the greater choice alone encourages closer attention to feeding methods. On the shelves of any pet superstore you can find products for a great variety of health, age, and disease conditions—often requiring one to choose between tartar control, hairball management, weight control, and so forth, as defining characteristics of the food. Such choices—and the reaction of the wild feeders against them—are all part of the "therapeutic ethos" that has prevailed in U.S. consumer culture since the early 20th century (Lears 1983). For some, the wild has come to be seen as a sort of therapy for problems ensuing from domestication, including those brought about by the pet food industry. There has been a similar reconsideration of the wild in thinking about human diets, as seen in the attention devoted to "Paleolithic diets" (Cordain 2002, Audette 1999). These trends result from similar situations, but my sense has been that wild feeders are not all that likely to be wild *eaters*. Consumers' identities and choices are chaotic and often contradictory (Fine 2006: 293).

A common claim made by wild feeders is that domestication matters little when it comes to the basic physiology and dietary needs of domestic "carnivores" (there is disagreement about the extent to which dogs might be considered "omnivores" with reports from archaeologists, anthropologists, and biologists providing support for both sides). Frequently citing the work of evolutionary biologist Robert Wayne (e.g., Wayne and Vilà 2003) to emphasize the genetic similarity between dogs and wolves (and discounting the work of biologist Ray Coppinger [2001], who emphasizes morphological and behavioral differences), proponents argue that no matter the breed, in matters of diet, a dog should be seen as essentially a wolf; the only remaining question being that of how to replicate such a diet using available materials. In pursuit of information on the eating habits of wolves and other wild canids, wild feeders study biologist David Mech's studies of wolves and report on trips to wolf sanctuaries, National Parks such as Yellowstone, and zoos. In 2004, Mech provided a daylong seminar to wild feeders, at which he presented all that he knew about the diet and eating habits of wolves, although he declined to discuss any application to dog feeding.

At a seminar in Atlanta in 2002, Australian veterinarian Tom Lonsdale instructed participants that we should consider our homes "zoos without bars" and act as "responsible zookeepers." Lonsdale characterized Hill's Science Diet as "a complete misappropriation of the word science" and part of a "distorted, corrupted culture." He got his audience laughing about the elaborate testing devices used by the pet food industry, including Waltham's (the Mars corporation's) "Feces Scoring Test." Participants, nearly all previously familiar with Lonsdale's basic ideas, accepted and disagreed with various aspects of the presentation, but "the wild" seemed generally to be an appealing orientation. For those new to Lonsdale's methods (or those of Billinghurst or other raw feeders), novelty may have been part of the appeal. One woman mentioned to me that feeding her dogs chicken carcasses "was like something on the Discovery Channel." The audience was diverse, however, and included another anthropologist (there not to investigate the wild feeders but ways of feeding an 18-year-old dog), a golf professional with chronically ill cats at home, a geneticist from the National Institute of Health contemplating a second career as a consultant for people interested in alternatives to commercial pet food, and women involved in animal rescue and breeding. Over lunch there was a lot of laughter as a group of us, including a couple of vegans, discussed the feeding of various animal body parts and carcasses and a sense of how strange and repulsive this would be to most people. As a group, we did not find much else in common.

Despite their diversity, the wild feeders tend to share a sense that their feeding choices are made in opposition to the power wielded by corporations. Meat, eating, animals—all of these have a long history of being rich in multiple meanings, of course, and associated not just with notions of the wild and the tame but also gender, civilization, nature, and culture. But among the wild feeders, a feeling of opposing corporate manipulation is especially powerful. That is not to say there is explicit agreement about this and what constitutes manipulation or exploitation. Billinghurst, for example, with his commercial aspirations and courting of industry giants, is often denounced as a "sell-out" by people who once found his books revolutionary, but that sentiment is not universal.

In *Dominance and Affection: The Making of Pets,* Yi-Fu Tuan writes that "domestication means domination: the two words have the same root sense of mastery over another being—of bringing into one's house or domain" (1984: 99). Both domestication and meaning may be more complicated matters than Tuan acknowledged, but certainly appreciating the wild can sometimes feel like escaping control.

Feeding the Hounds

Before closing, I would like to consider a case in which the methods of feeding domestic dogs advocated by the wild feeders discussed above are not considered "wild." Despite the processes of globalization preoccupying the participants at the Petfood Forum, perceptions of the wild and its opposites can still be rooted in very particular local circumstances.

In May 2002, I crouched beside a man with the title of "Master of Foxhounds" in the yard of a kennel in Shropshire, England. We peered between iron bars at a white foxhound bitch nursing half a dozen pups on a straw bed. "The foxhound," the Master said very quietly, reaching in to stroke her chin, "is the closest thing Britain has to a wolf." For a moment, I was not sure I had heard correctly. I found the remark startling for several reasons. Most immediate was the orderliness of the kennel and the hounds. I had just come from watching the "hunt kennelman" lead a group of some 35 young hounds around country roads on a bicycle, the pack milling around behind while waiting for trucks to pass. But probably even more accounting for my surprise was my long-standing and largely unconscious acceptance of foxhounds and foxhunting as representing English "culture." It was only later, when I considered the remark's larger context, that the foxhound–wolf comparison seemed less surprising.

That summer of 2002, Parliament was considering a ban on hunting with dogs, a ban promoted for years by animal rights activists. Farmers were struggling to abide by a host of new regulations aimed at thwarting a resurgence of foot-and-mouth disease and at eradicating BSE (bovine spongiform encephalopathy, sometimes known as "mad cow disease"). I also heard a lot of talk of city people buying houses in the country. So it should not have been surprising when the Master of Foxhounds and his colleagues depicted, not just the hounds in their care, but also themselves, as in some ways belonging to the ranks of endangered species. "You can't do anything in this country anymore," said another Master of Foxhounds on the same day I visited the kennel.[6] After a discussion of her efforts to halt the proposed ban on foxhunting, her statement seemed to reflect frustration with a shift in sensibilities and with the expansion of state and other centralized authority over the countryside. Similar concerns likely influenced her colleague's remarks, during my visit to the kennel, that "the foxhound" is something "Britain has," similar to how "North America has the wolf," as he put it, and his suggestion that foxhounds are less domesticated than other dogs.

Before I traveled to Shropshire, anthropologist Garry Marvin had explained to the Master of Foxhounds and to the hunt kennelman that I was interested in the practice of feeding foxhounds "deadstock" (also known as "fallen stock") or the carcasses of cattle and sheep that had died on surrounding farms. At the American Anthropological Association Annual Meeting the previous year, Garry had mentioned to me that this was the only way the foxhounds were fed. I had heard similar reports from others experienced with English foxhunting, although I had also heard reports of foxhounds being fed leftovers from a fish and chips shop. But during my visit to their kennel, the hounds' caretakers explained that recently they had been feeding a combination of deadstock and commercially produced dry kibble. Dead livestock had been in shorter supply as a result of the recent foot-and-mouth crisis and concerns about BSE (although dogs have not been shown to be susceptible to BSE, I was told that the hunt was prohibited from feeding cows older than 24 months, the age at which they were thought to become potentially capable of transmitting the disease). So, for my visit, the men had saved up a number of carcasses, some 10 calves and a couple of lambs, that ordinarily they would feed soon after their delivery (see fig. 11.2). They would supplement the carcasses with dry food that came in bags and that I saw stacked several feet high in a small storage building.

I also saw the carcasses of the calves and lambs lined up in preparation for feeding, heads and hides removed, the hides to be sold by the hunt's employees as a way of supplementing their wages. Garry and I stood in the "flesh house" as two groups of about 35 hounds each, the hounds and the bitches, were let in to demolish the carcasses. It took about 15 minutes for a group to reduce carcasses weighing nearly 100 pounds each to a few large fragments of pelvic bones and femurs. Then a bag of dry kibble was spread on the concrete floor for the hounds to clean up while the hunt kennel man put the remaining bone fragments into an incinerator. After they ate, the hounds were let out to run in a large field of sheep (where they seemed to enjoy another meal, this one of sheep droppings).

While we were visiting the kennel, a small car drove up with the legs of a dead calf protruding from the back of the car. Garry Marvin and I were told that it was a local farmer dropping off the calf. The Master of Foxhounds explained that he and the hunt's employees used to spend many hours traveling around the countryside, collecting dead livestock, as well as killing injured or diseased animals. He felt fortunate that they had been able to switch to the present system, one in which the farmers

Figure 11.2. Foxhounds eating carcasses of livestock. (Photograph by author, 2002)

pay the hunt for accepting the carcasses they bring to the kennel, a shift made possible in part by tighter government regulations regarding disposal (I later discussed this situation with people who had been involved with hunts in the past and they were amazed to hear about the convenience of having farmers deliver carcasses to the hunt). Even beyond feeding the hounds, the hunt had reason to provide farmers with the service of disposal at low or no cost, because the hunt depended on farmers for permission to hunt on their land, permission that had to be granted prior to each hunt. Feeders of the hunt's hounds could also supplement their wages by selling hides. Thus, feeding practices were enmeshed in the local economy.

The men I spoke with at the kennel said they were surprised to hear that anyone would find their feeding practices controversial, apart from possible concerns about BSE. They also appeared unfamiliar with the notion of dog food companies as manipulative, powerful, and greedy. Their primary concern seemed to be the power of government. As soon as I arrived I was told that although they were happy to meet with me and let me observe them feeding the hounds, they did not want to be identified in anything I would write. Any photographs that I might take should be "deniable." I was told that this was not because they were doing anything necessarily illegal, but there were so many regulations, changing all the time, and there was always a possibility that they were not following the current rules exactly.

Although he spoke reverently of foxhunting as part of the traditional life of the countryside, the Master of Foxhounds seemed to take pride in being a modernizer. Garry Marvin confirmed this impression, and Garry had worked with him for a number of years while doing ethnographic research on foxhunting. The Master explained that he would prefer to switch the hounds over entirely to commercially produced dry food ("kibble"), in part because of recent difficulties obtaining an adequate number of carcasses, but also because he thought that as long as he could feed a brand with a high enough protein percentage (27%), he felt the hounds would "do best," a performance he attributed to "vitamins" added to the food. He thought they managed quite well eating the carcasses however, far better than if he had to feed them a less-than-premium brand. In our discussion of feeding, the likeness to wolves that he had mentioned earlier did not come up, or any mention of the hounds needing to be fed differently from other canines for any reason other than the physical demands of hunting. At any rate, so far the hunt could not afford to feed a commercial product exclusively and he did not expect that they would be able to do so any time in the foreseeable future.

After visiting the kennel, I learned that the Countryside Alliance, a group formed in opposition to the proposed restrictions on hunting, featured "deadstock collection" on its Web site, on a page devoted to hunting's "contribution to the rural community," with a photograph of what is identified as a hunt kennelman skinning a dead calf (Countryside Alliance 2003). The alliance states that in disposing of dead farm animals the hunt fills a basic need for farmers. "Approximately 400,000 carcasses a year are disposed of by hunts, free of charge or at much reduced rates," the site reported, adding:

> The huntsmen and other kennel staff take on the responsibility of butchering the carcass in the correct manner to be able to dispose of offal as now required under regulations concerning BSE [The Specified Risk Material Order and Regulations, 1997]. The bulk of the carcass is fed to the hounds.

An article appearing in the *New York Times* following a march in London organized by the alliance quoted a British farmer's gratitude for this useful service (Lyall 2002). No mention was made, in the *New York Times* article or on the Countryside Alliance web site, of "the wild" or of how wolves might eat.[7]

Feeding and Domestication

In the last decade of the 20th century, zoos began to cease offering carnivore-feeding spectacles. There are a variety of reasons, and one is that many zoos, at least in North America, now feed exclusively dry kibble from bags, purchased from Nestlé and other companies (Houts 1999).[8] When feeding spectacles at zoos were still fairly common, Yi-Fu Tuan attributed their popularity to an ability to make humans feel civilized and superior to animals (Tuan 1984: 80; see also Malamud 1998: 234). However, as I have tried to show here, perceptions of animal feeding, the wild and domesticated, the human and the animal, are much more varied, context dependent, and unstable than such accounts acknowledge.

As anthropologists and other scholars continue to rethink domestication, we should keep in mind that people are reconsidering domestication in a great variety of contexts and with very different concerns and perspectives. As far as people's abilities to influence others' understandings and produce authoritative narratives, of course it is not an equal playing field and some people (e.g., anthropologists, biologists, and advertisers?) wield more power and influence than others. But just as there are unforeseen consequences of domestication, there are also unintended consequences of domestication's associated technologies, commodities, and industries.

Am I suggesting that animal feeding industries are themselves an engine of "domestication"—of humans, animals, or both? I am not inclined to call them that. I have focused here on investigating how people use abstract concepts like domestication, rather than on using them in a very direct way myself. Part of my reluctance about calling anything domestication stems simply from the fact that it is confusing to be studying how other people use such abstractions in relation to particular contexts while at the same time employing them descriptively. Also, domestication is, like culture, an abstraction with particularly fluid and contested meanings; I tend to prefer greater precision when possible.

However, I do think it is worth considering the feeding of animals in relation to domestication as a concept, especially in relation to the questions about domestication raised by other authors in this volume. For example, Nerissa Russell proposes that we consider animal domestication as a form of kinship. In the case of pet food, we see an industry promoting the view of pets as family members, although the consequences are not entirely controlled by (or necessarily in the interests of) that industry.

Helen Leach also emphasizes unforeseen consequences and how it is possible to see humans as "domesticated" along with other species. Peter Wilson, Gillian Feeley-Harnik, Marianne Lien, and Agustin Fuentes encourage us to think not just about relationships among species but also how these relationships are configured spatially, how they are affected by spatial arrangements, built environments, and by practices other than the controlled breeding of animals and plants. The cases I have explored here involving animal feeding show human, animal, and plant species intertwined and in ways that lead to unexpected outcomes (e.g., the shift from pet to companion animal leading to caretakers of companion animals raising questions about species-appropriate diets, the notion that not just humans but also their animals have an "inner beast"). They show that understandings of domestication and the wild are tied to particular spaces (the laboratory, the field, the kitchen, the pet superstore, or the farm) and also to technologies of packaging, storage and processing (pull-off tops on containers, freezers, and the meat packing and grain industries). If we intend to consider domestication an ongoing process, we should pay attention to how life is reproduced through feeding as well as breeding, and we should pay attention not just to the now traditional question of the relationship between domestication and property but also to that between domestication and expanding consumer capitalism.

Notes

Acknowledgments. Research for this chapter was supported by grants from the Hewlett Mellon Fund for Faculty Development at Albion College. The author is grateful to Garry Marvin for making it possible to observe the feeding of foxhounds at a kennel in England and to Laura Waldo, Alison Tyler, Ian

Billinghurst, Tom Lonsdale, and Lew Olson for sharing their perspectives on companion animal feeding methods. Many thanks also to Susan Hegeman, Rebecca Cassidy, Sarah Franklin, and two anonymous readers for especially helpful responses to earlier versions.

1. Richard Bulliet (2005) argues that the popularity of horror films and of wildlife documentaries is an aspect of what he calls "post-domesticity," an era marked by fascination with animals and the wild as well as discomfort with their domestication and exploitation.

2. For North Americans the shift away from the overt marketing of rabbit and horse meat as "pet food" was possibly hastened also by the rise of grain-based pet food and the sales of dry food or "kibble" sold in bags. Although especially after the end of WWII meat rationing, dry food for dogs and eventually cats was still likely to contain meat and other by-products of the meat packing industry, it was probably easier for feeders to think of it more generically as "food."

3. As Katherine Grier notes in her history, *Pets in America*, "As part of the American food industry, the dog and cat food business has undergone the same kinds of historical processes, including corporate consolidation and the elimination of many small, regional brands, that have shaped packaged foods for people" but she also notes that the Internet has led the resurgence of small "boutique" companies or "niche producers" (2006: 290–291). For an industry perspective on the "top 10 global petfood players," see Kvamme 2006.

4. It was not uncommon in the mid–20th century for pet food companies to mention laboratory research animals in publications. In 1935, the manufacturers of Friskies cat food published pictures of laboratories where they reported their food had been tested on "five generations of Albino rats" (Grier 2006: 287).

5. I use the terms "wild feeders" and "wild feeding" when discussing feeding methods based in some way on the eating habits of wild and feral animals. People using such methods use a variety of other terms, including "raw feeding" and "whole carcass" or "prey model feeding."

6. This must be a common way of expressing nostalgia in the United Kingdom. In her study of horseracing (and kinship and class), Rebecca Cassidy includes almost the exact same quotation (2002: 46).

7. In the press coverage of the fallen stock service provided by hunts and discussions of the practice in the British Parliament (the practice of hunts disposing of livestock carcasses has often come up in relation to the economic impact of the ban on hunting with dogs), it has been considered unnecessary to explain how the carcasses were disposed of by the hounds. Yet when I have discussed such feeding practices with U.S. audiences, including veterinarians,

people have responded with great surprise, confusion, and with many questions about how and whether it is possible for dogs to eat the carcasses of large livestock.

8. Although carnivore feeding spectacles at North American zoos generally have been abandoned, zookeepers are also questioning the practice of feeding carnivores commercially produced dry food and zoos that had ceased feeding carcasses are now reviving the practice (Derr 2003; Houts 1999).

References

Anderson, Kay. 1998. Animal domestication in geographic perspective. *Society and Animals* 6 (2): 119–134.

Audette, Ray. 1999. *NeanderThin: Eat like a caveman to achieve a lean, strong, healthy body.* New York: St. Martin's Press.

Bulliet, R. 2005. *Hunters, herders, and hamburgers: The past and future of human–animal relationships.* New York: Columbia University Press.

Cassidy, Rebecca. 2002. *The sport of kings: Kinship, class and thoroughbred breeding in Newmarket.* Cambridge: Cambridge University Press.

Chappie, Ltd. n.d. *Cats, dogs, and diets: A scientific approach.* London: The Kennel Club Library.

Coppinger, Raymond. 2001. *Dogs: A startling new understanding of canine origin, behavior and evolution.* New York: Scribner.

Cordain, Loren. 2002. *The Paleo diet.* Hoboken, NJ: Wiley.

Countryside Alliance. 2003. *Home page.* Available at: http://www.countryside-alliance.org/edu/edu2-5-3.htm, accessed January 15.

Crossley, Adrienne. 2003. *The future of the pet food industry.* Paper presented to the Petfood Forum, Chicago, Illinois, April 1.

Derr, Mark. 2003. Zoos are too small for some species, biologists report. *New York Times,* October 1. (Internet edition)

Devaney, Valeri. 2003. "Feeding dogs carbs, was I'm on this list and I can't get off." July 20. Available at: Altvetskept-l, Altvetskept-l@po.missouri.edu mailing list.

Fine, Ben. 2006. Addressing the consumer. In *The making of the consumer: Knowledge, power and identity in the modern world,* edited by Frank Trentmann, 291–311. Oxford: Berg.

Grier, Katherine C. 200. *Pets in America: A history.* Chapel Hill: University of North Carolina Press.

Haraway, Donna. 2003. *The companion species manifesto: Dogs, people, and significant otherness.* Chicago: Prickly Paradigm Press.

Houts, Lee. 1999. Supplemental carcass feeding for zoo carnivores. *The Shape of Enrichment* 8 (1): 1–3.

Kemper, Steven. 2001. *Buying and believing: Sri Lankan advertising and consumers in a transnational world.* Chicago: University of Chicago Press.

Knight, John. 2005. Feeding Mr Monkey: Cross-species food "exchange" in Japanese monkey parks. In *Animals in person: Cultural perspectives on human–animal intimacies,* edited by John Knight, 231–253. Oxford: Berg.

Kvamme, Jennifer. 2006. Top 10: Profiles of pet food leaders. *Petfood Industry* 48 (1): 6–15.

Leach, Edmund. 1964. Anthropological aspects of language: Animal categories and verbal abuse. In *New directions in the study of language,* edited by Eric H. Lenneberg, 23–63. Cambridge, MA: MIT Press.

Lears, Jackson. 1983. From salvation to self-realization: Advertising and the therapeutic roots of the consumer culture, 1880–1930. In *The culture of consumption: Critical essays in American history, 1880–1980,* edited by Richard Wrightman Fox and T. J. Jackson Lears, 3–38. New York: Pantheon Books.

——. 1994. *Fables of abundance: A cultural history of advertising in America.* New York: Basic Books.

Lévi-Strauss, Claude. 1971. *L'homme nu. Mythologiques IV.* Paris: Plon.

Lyall, Sarah. 2002. Grumbles grow louder in quiet rural Britain. *New York Times,* October 2. (Internet edition)

Malamud, Randy. 1998. *Reading zoos: Representations of animals and captivity.* New York: New York University Press.

Mieszkowski, Katherine. 2006. *The beef over pet food.* Salon.com, January 19. Available at: http://www.salon.com/news/feature/2006/01/19/raw/index_np.html.

Milliet, Jacqueline. 2002. A comparative study of women's activities in the domestication of animals. In *Animals in human histories: The mirror of nature and culture,* edited by M. J. Henninger-Voss. Rochester: University of Rochester Press.

Ralston Purina Company. 1937. *Your dog: His care and training.* St. Louis: Purina Mills.

Ramey, David. 2003. "Rewrite." June 25. Available at: Altvetskept-l, Altvetskept-l@po.missouri.edu mailing list.

Russell, Nerissa. 2002. The wild side of animal domestication. *Society and Animals* 10 (3): 285–302.

Sax, Boria. 2000. *Animals and the Third Reich: Pets, scapegoats, and the Holocaust.* New York: Continuum.

Sendak, Maurice. 1963. *Where the wild things are.* New York: Harper and Row.

Thomas, Keith. 1983. *Man and the natural world. Changing attitudes in England, 1500–1800.* New York: Pantheon.

Thoroughbred Dog Biscuit. n.d. Promotional leaflet in the author's possession.

Tuan, Yi-Fu. 1984. *Dominance and affection: The making of pets.* New Haven, CT: Yale University Press.

TV Guide. 2001. Television Schedule, October 13–19.

Vidas, Anath Ariel de. 2002. A dog's life among the Teenek Indians (Mexico): Animals' participation in the classification of self and other. *Journal of the Royal Anthropological Institute* 8: 531–550.

Wayne, Robert K., and Carles Vilà. 2003. Molecular genetic studies of wolves. In *Wolves: Behavior, ecology, and conservation,* edited by L. D. Mech and L. Boitani, 218–238. Chicago: University of Chicago Press.

White, David Gordon. 1991. *Myths of the dog-man.* Chicago: University of Chicago Press.

Wigley, Mark. 1999. The electric lawn. In *The American Lawn,* edited by George Teyssot, 154–195. New York: Princeton Architectural Press.

Index